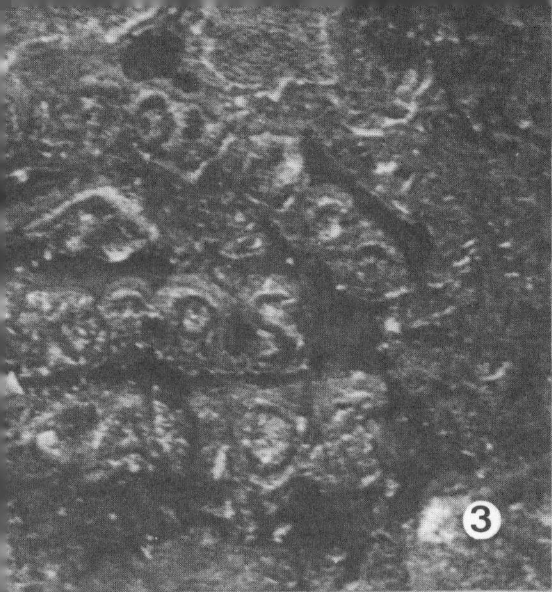
3

6

4

7

5

8

ADELAIDE LACE

PUBLISHED AS A LIMITED EDITION
OF 1,001 COPIES SIGNED BY THE AUTHOR,
OF WHICH THIS IS NUMBER

190

E. Graeme Robertson

Also by E. Graeme Robertson
(Published by Georgian House)

Victorian Heritage
Sydney Lace
Early Houses of Northern Tasmania
(with Edith N. Craig)
Ornamental Cast Iron in Melbourne
Early Buildings of Southern Tasmania
Early Colonial Furniture
(with Clifford Craig and Kevin Fahy)

ADELAIDE LACE

E. GRAEME ROBERTSON

RIGBY

RIGBY LIMITED • ADELAIDE • SYDNEY
MELBOURNE • BRISBANE • PERTH

First published 1973

Library of Congress Catalog Card Number 77-183436
National Library of Australia Card Number and ISBN 0 85179 623 0

Monophoto typeset in Australia by Modgraphic Pty Ltd, Adelaide
Printed by Macarthur Press Pty Limited, Sydney
Bound by Stanley Owens & Sons Pty Ltd, Sydney
Registered at the G.P.O. Adelaide for transmission by post as a book

RIGBY LIMITED, 30 NORTH TERRACE, KENT TOWN, SOUTH AUSTRALIA
Adelaide • Sydney • Melbourne
Brisbane • Perth

TO MY WIFE

For whom my books are an inadequate memorial.
Her qualities made their writing possible.

Don't forget to speak scornfully of the Victorian Age, there will be time for meekness when you try to better it.

J. M. Barrie

The Victorian house is like a rugged character who steals the scene from the smooth-faced leading man; rich in shape, colour and texture; it has long been a favourite model for artists and photographers.

John Maass[1]

CONTENTS

ACKNOWLEDGMENTS

IT GIVES ME MUCH pleasure to acknowledge the help given by the following: Dr Richard Rischbieth of Adelaide for enthusiastic help in the search for ornamented buildings.

Mr J. Magnussen of the Griffin Press sought answers to questions posed from Melbourne, which resulted in his photographing many of the manufacturers' names which form the end-papers. He took the photographs shown in plates 108 and 159.

The following people provided invaluable material for this book, and, although I am sure that the donors will feel recompensed by finding the material used, I can never adequately express my gratitude.

Mr Archie Forwood of Adelaide presented me with a fine copy of the second edition of the catalogue of the "Sun" Foundry of Adelaide, then conducted by A. C. Harley and Company. Plates 2–21 are taken from this catalogue. Mr Noel Goss was so kind as to give me the sixth edition of the catalogue of Macfarlane's Saracen Foundry of Glasgow, which was valuable for comparison. The late Sir William Johnston gave me a copy of the *Illustrated and Descriptive Catalogue of Ornamental Cast and Wrought Iron and Wire Work, manufactured by Charles D. Young and Company*.

Mr L. A. Daniel sent me a copy of the catalogue of Metters Foundry, Adelaide (plates 182–187). Dr Donald Simpson, unable to provide records of a pioneer foundry conducted by his forebears, effected the introduction. Mr Daniel also provided the information of Metters Limited, included in the text.

Catalogues of ornamental castings made in Australia, and indeed throughout the world, are rare items. I have been able to purchase L. N. Cottingham's *The Smith and Founder's Director* and a number of catalogues. These include the fifth edition of Macfarlane's catalogue, *Examples Book of Macfarlane's Castings,* and the catalogues of Barbezat et Cie and of the Sun foundry of Geo. Smith referred to in the text.

Mr Malcolm Higgs in Edinburgh bridged the gap in our knowledge of Corio Villa, Geelong.

Miss P. Reynolds, Archivist of the La Trobe Library, State Library of Victoria, identified "the late Mr Gray, Colonial Land Commissioner at Geelong" who ordered Corio Villa. Mr P. L. Brown, past president of the Geelong Historical Society, added to the information.

Mr Gary Kay of Antique Decor pointed out registration marks on English castings and supplied the *Antique Dealers and Collectors Guide* extracted in the text.

The following libraries made valuable contributions:

The Baillieu Library, University of Melbourne, holds a portfolio entitled *Hemming's Patent Improved Portable Houses,* some of the illustrations of which are reproduced in this text. Miss M. Lugton very pleasantly provided excellent liaison with other libraries.

The Public Library of South Australia kindly provided excellent photocopies of the second edition of the *Illustrated Catalogue of Fulton's Castings,* some of which are reproduced in plates 22–26 and 95–99.

The Library of New South Wales possesses three other catalogues of Walter Macfarlane's castings (Vol. II of the sixth edition of the *Illustrated Catalogue of Macfarlane's Castings, Illustrated Examples of Macfarlane's Architectural Iron Work,* and recently Miss J. Hunter, Reference Librarian, drew my attention to a new acquisition *Examples Book of Macfarlane's Castings*).

I am indeed grateful to Sir Edward Morgan and Mr S. H. Gilbert for allowing me to share with my readers the fruits of their researches, and to the Oxford University Press, Melbourne, for permitting me to quote *verbatim* from their magnificent book, *Early Adelaide Architecture.*

Mr Mark Strizic processed, with the extraordinary high standard which he has set, the remainder of the photographs, which were taken by the writer.

Miss Joan Robertson has, with unique critical and perfectionist ability, typed the manuscript and ensured uniformity of presentation throughout.

INTRODUCTION

THIS BOOK IS DESIGNED as a supplement to *Victorian Heritage*[2], *Sydney Lace*[3], and *Ornamental Cast Iron in Melbourne*[4], to extend the record of the usage of decorative castings in Australia and to show the fascination of Adelaide in this regard. Two comparison volumes are to follow: *Ornamental Cast Iron in Queensland,* and an intensive study of one suburb, that of Carlton in Victoria. Each book is complete in itself, and together they illustrate the diversity of architecture and the manner of usage of cast iron in cities in one continent.

The subject is peculiarly suited to photographic presentation.

Andreas Feininger wrote[5]:

> One can photograph effectively only those subjects which capture his interest. Interest is essential because a photograph is basically a means of communication—the sharing of information, experience or feeling with others. To be successful as a form of communication, a photograph must therefore capture the observer's interest. This is only possible if, to begin with, a photograph is charged with the photographer's own interest in the subject he depicts. To be able to really "see" a subject and capture its essence in a photograph, one must establish a true contact with the subject through study, understanding or immediate intense response. And this, of course, is possible only if the driving force comes from interest.
>
> In producing the preceding photographs, I made a great effort to use the camera as an extension of my eye.

William Mortensen defines his formula for photographic success as issuing *The Command to Look*[6].

Bill Brandt, in 1948, wrote[7]:

> The photographer must have, and keep in him, something of the receptiveness of the child who looks at the world for the first time . . .

> What we see is often what our prejudices tell us to expect to see, or what our past experiences tell us should be seen, or what our desires want us to see. Very rarely are we able to free our minds of thought and emotions and just see for the simple pleasure of seeing. And so long as we fail to do this, so long will the essence of things be hidden from us.

I myself have found satisfactory those photographs in which the camera clearly speaks its own language—which, in the context of this book, is most often the language of realism, in which the object or idea in the photographer's mind is isolated, clearly demonstrated, tells its story, and commands attention.

Cities may be perceived in many ways, and so may also their component houses. Seeing the whole often makes us unaware of details, while seeing details may add immeasurably to our appreciation of the whole. So it is with cast iron decoration, which in our cities was so common as to be passed by without being seen, and indeed until recently was regarded as so old-fashioned as to be unworthy of notice. So the destruction of one of our greatest assets passed unnoticed, and even now town councils and other authorities plan and effect its destruction without understanding the significance of their acts.

The arousal of the citizenry to effective protest is the one way in which the common man can stop this fiesta of destruction. It is hoped that this book will help in bringing awareness of our heritage, and inhibit the vandalism of those in office, who, lacking aesthetic appreciation, think neither of the past nor of the future as they wreak destruction. This potential annihilation is as culpable, even though the motive be different, as the infinitely smaller acts of vandalism done by the mentally destitute or the rebellious youths of our community.

It has been truly said that one hundred years pass before a style becomes evaluated, restudied, and appreciated. As an example of this *The Australian Encyclopaedia* of 1927 may be quoted[8]:

> Towards the middle of the 19th century, when taste was at a low ebb and cheap ornament popular, commercial firms made a speciality of building accessories such as verandah pillars, railings and roof crestings, in cast iron of the poorest design, and the simple Italian style of domestic architecture degenerated into an over-ornamented type, of which many examples still exist.
>
> . . . Fortunately commercial architecture largely escaped the craze for cast iron ornamentation which was so prejudicial to our domestic architecture. In some instances a tendency to over-richness in ornament is noticeable; but on the whole reticence, dignity and good proportions distinguish the more permanent examples.

It is fortunate that so much cast iron decoration has withstood such onslaughts for so long—especially when it was regarded as old-fashioned and only fit for demolition or substitution. Unfortunately, the destruction of shop verandahs by decree has been an infectious pandemic afflicting city and town councils. The melancholy results may be seen in many of our towns and cities. The buildings were designed with the verandah as an integral part: to remove the verandah destroys the proportions of the building, and is an act of defacement. The shorn appearance of some of our country towns stands as a memorial to those who have transformed them from pleasant havens to repellent locations. Towns which were attractive reminders of our past, which would have gathered added virtue with the passing years, become places in which the traveller sees no reason to halt, and indeed, hurries through with relief after passing.

The shop verandah is, curiously, aesthetically satisfying, and is a functional unit with the building, shielding the passer-by from the hot summer's sun and from rain, providing a pleasant place for greeting friends, while shading shop windows and keeping the interior cool. "Modernisation" is usually ephemeral, transforming a heritage which has stood the test of time into something characterless, and even downright ugly. Replacement by cantilever awnings, or, worse still, by telescopic multi-colour metal shades, further mutilates the building, cutting it horizontally in half by an utterly foreign structure. If authorities are not too stubborn, or too obsessed with the heady intoxication of their destructive power, they should compare cities and towns where iron verandahs remain with those denuded.

The problem of destruction or preservation of individual houses may be solved by immediate financial considerations, or may depend upon the owner's failure to recognise that he possesses an historical treasure which will become more valuable as time passes. Alteration of the verandah may be as destructive as demolition. Replacement of a balustrade by wrought iron or by plywood, and, worse still, the filling-in of the whole front, destroys the attractiveness of the house, and maybe of a whole terrace, and even of a street. A coat of paint almost miraculously transforms the appearance of a cast iron-bearing house. A previously dull and seemingly uninteresting house becomes an eye-catcher.

Modern conditions certainly provide compelling reasons for the destruction of some buildings, but wanton and wholesale destruction is to be deprecated. The National Trust has provided standards of evaluation of older buildings, but has insufficient power to enforce preservation. Preoccupation with older buildings may turn attention from this younger, but equally significant and more unique, form of our indigenous architecture.

Fortunately, the forces of preservation are gaining ground. The greatest force is the desire of citizens to own and live in these historical, comfortable, and well-constructed buildings. A few are still to be found in the centre of cities, and many remain in the nearby suburbs which were built up in the latter part of last century. A renaissance of these areas, once regarded as poor, is under way, but many are battling against authorities who blindly destroy after stigmatising them as slums.

The people who live in these houses are a happy breed, and very proud of their possessions. Retention of whole neighbourhoods is still more desirable, and is being effected in many cities of the world. It is time that we Australians should follow good examples elsewhere with the same avidity as we follow the bad qualities of "affluence."

Roger Aldridge, in *The Age* of 22 November 1969, provided a humorous example of "strine," and a deeply serious protest against destruction, in an article entitled "The Rape of our Gloria Seritage."

And so it goes on, this senseless, insensitive destruction of our past . . .

There is still sufficient cast iron ornamentation to add to Australia's reputation for tourism, in which the capital cities would each provide stopping places, attracting many visitors from abroad.

AUTHOR'S NOTE

In the legends the following abbreviations will be found in the documentation which is placed in parenthesis:

H. C.—Harley's Catalogue (second edition)
F. C.—Fulton's Catalogue (second edition)
No.—number of item in catalogue or registration number
p., pp.—page, pages
plate—refers to illustrations in *this* volume, unless otherwise indicated
N.S.W., Vic., S.A., and Qld. indicate the States in which designs were registered. Documentation is usually placed in brackets throughout this book, so that it may be avoided by the general reader if he so desires.

Throughout the book the word "verandah" is used to denote a roofed structure built out from the house at entrance floor level, and the word "balcony" to denote a roofed or unroofed platform with access from higher floors, possibly above a verandah.

I

ORNAMENTAL CAST IRON IN ADELAIDE

ADELAIDE IS A CORNUCOPIA of rich, delicate cast iron ornamentation, and is also a very beautiful city. The cast iron ornamentation is much less prolific than in Melbourne, but it has many variations. Melbourne is losing its individuality, and soon, in spite of the efforts of the National Trust, only a few early buildings and the magnificent parks will remain to attract visitors. Sydney will suffer less, because of its setting in the harbour, and the excitement which prevails in the areas where the inhabitants have forced preservation. Elsewhere in the world cast iron bearing areas are being preserved, but until preservation becomes a matter of parliamentary enactment, this is not likely to happen in Australia. Adelaide still retains its old charm, in spite of some changes, and it is to be hoped that the future generations of citizens and visitors will still be able to enjoy that charm. Indeed it will become more remarkable and valuable as the years pass.

Adelaide may well lay claim to being the best-planned State capital in Australia —its present form being determined by the plan prepared by Colonel Light, Surveyor-General. The city, named Adelaide at the request of King William IV in honour of his Queen, was divided into North and South Adelaide. The two parts were surrounded by a green belt, and were separated by a band of open country, about half a mile in breadth, through which flows the Torrens River.

South Adelaide, now usually referred to as Adelaide, was planned on level ground south of the Torrens, approximately one square mile in extent, bounded by wide roadways named North, South, East, and West Terraces. The area contained five squares, the central one being called Victoria Square. The rest was subdivided geometrically in grid fashion. This area became the commercial part of the city.

North Adelaide rises from the river flats to higher ground, and has remained largely residential.

The city developed during the cast iron era, and fortunately much of the ironwork remains. In South Australia, as in other States, hotels tended to use cast iron decoration, and they have proved to be splendid protectors of this form of architectural design. The Corporation of the City of Adelaide is to be congratulated upon preserving some of the distinctive atmosphere of their city.* Much more may be hoped for in the future.

IMPORTANCE OF IRON-FOUNDERS

The part played by the iron-founders in our architecture has not been adequately recognised. In *Victorian Heritage*[2] two fallacies are discussed; one, the common tendency to refer to cast iron as wrought iron, and the other that our cast iron decoration was brought as ballast by ships sailing to Australia. Both fallacies persist. The matter will be dealt with in some detail in the final section of this book; but it is sufficient to state here that most of our cast iron ornamentation was produced locally, although castings manufactured, or designed, in Great Britain have been seen in Melbourne[2, 4], Sydney[3], and Tasmania[9]. The names of eighty-seven Australian firms producing ornamental architectural castings in the latter part of the nineteenth century were listed in *Victorian Heritage*[2]. Undoubtedly many more existed, for example, nine further factories in Victoria are known to have produced cast iron tables, chairs, and ventilator fronts, and probably also made other ornamental castings. A number of country towns were supplied with ornamental ironwork by local foundries.

Thus iron-founders played an historical role which they little suspected. Trade catalogues are important historical documents, but unfortunately were disposed of in the manner of most catalogues, so that few survive. The *Illustrated Catalogue* of William Stephens's Excelsior Foundry, South Melbourne[10] was illustrated in its entirety in *Ornamental Cast Iron in Melbourne*[11]. Adelaide was the site of a number of outstanding foundries, and fortunately catalogues of three of these have been located: The "Sun" Foundry, Fulton's Foundry, and Metters' Foundry.

CATALOGUES OF THE "SUN" FOUNDRY, ADELAIDE

Two editions of the catalogue of the "Sun" Foundry were published—the first in 1897[12] and the second in 1914[13]. The first contained 108 pages, the second 137. The title pages are shown in plates 1 and 2.

At the time of the first edition Colin Stewart and Allan Cameron Harley were the proprietors, while by the time of the second edition, the firm had passed into

* An enquiry directed to Mr R. W. Arland, Town Clerk of Adelaide, elicited the following information: "Between September 1965 and April 1966 the Corporation widenened the footways of Rundle Street between King William and Pulteney Streets to permit a better pedestrian flow. Reconstruction of Hindley Street occurred between August 1969 and June 1970. These works required the previous removal of pillar verandahs, many of which, in Hindley Street, were extensively decorated with cast iron lace work. In Rundle Street the verandahs were nearly all wood structures with very little cast iron ornamentation.

The Council asked the National Trust of South Australia to recommend which verandahs should be permitted to remain because of their architectural or historical significance. The verandahs recommended for retention were permitted to remain, and in one case the owners of the Royal Admiral Hotel were permitted to retain the verandah, provided it was restored to its original Victorian appearance by using cast aluminium lace work on the pillars and friezes.

Apart from the above, it has not been found necessary to require the removal of pillar verandahs on such a large scale, and where one or two verandahs have been removed from a street these have no historical or architectural significance."

the hands of the Harley family. However, the Stewart and Harley families remained intertwined for some time. Many of our foundries, as might be expected, were founded by men who had been employed in similar work in the United Kingdom. I believe none of the original foundries survive under their original names. That the "Sun" Foundry was no exception is shown by information kindly provided by Mr Colin J. Short, grandson of Colin Stewart.

> My maternal grandfather, Colin Stewart, served his apprenticeship somewhere in Scotland with the Sun Foundry. On arrival in Australia he settled in Adelaide and began the business with Allan Cameron Harley in a small street off Hindley Street in Adelaide, at the rear of the South Australian Brewery.
>
> The partnership was dissolved before 1910 and the business was retained as A. C. Harley and Co., and eventually the son of the original Allan Cameron Harley, who was also named Allan Cameron Harley, took over the management of the foundry. This son married Agnes, the daughter of Colin Stewart, who had a son working in the foundry as a pattern maker for many years.*
>
> Eventually, about 1924, A. C. Harley and Co. were bought out by Forwood, Down and Co., then established in Hindley Street as moulders, boiler makers, and engineers. The "Sun" Foundry was transferred into Forwood, Down's workshop, and the Harley property sold to the Brewery. Sand patterns and boxes were moved to the moulding floor previously owned by the Fulton Foundry at Kilkenny. William Stewart, the son of old Colin, managed the foundry for some years.
>
> However, to reduce overheads in the depression, the Kilkenny Foundry was closed and the business was carried on in Hindley Street. In more recent times, Forwood, Down and Co. was bought out by Johns and Waygood, and became Forwood, Johns and Waygood, and then more recently still Johns and Waygood and Perry Engineering amalgamated. Thus, in a space of forty-odd years, all the major engineering works in Adelaide are now under the one control.

The second edition of the catalogue was published after the heyday of iron ornamentation in other States. The contents are indexed under the following headings: Lamps and Stairs; Standards; Rollers; Seats; Fountains; Plumbers' Requisites; Gratings; Wheels, Columns, Railings, and Gates; Brackets; Friezes and Fringes; Builders' Requisites; Blacksmiths' Requisites; Miscellaneous.

It is, of course, impossible to reproduce the large catalogue in this volume. The balusters are shown in plates 3–11. Although leaves, twigs, and flowers are frequent decorative motifs, little attempt was made at a naturalistic representation. Botanical authorities who have inspected the photographs used in this book find few recognisable specimens. Fern fronds are the most naturalistic, while daisies

* I have been able to find evidence of only one other pattern maker's name.

and sunflowers are readily recognisable. Other representations tend to be imaginative, or stylised. The latter follow the precedent set by the ornamental carvings of classical antiquity, as, for example, acanthus leaves, palm, and honeysuckle (anthemion) motifs.

Matching sets of friezes, fringes, and brackets were available (plates 12–14), although sometimes a builder chose to mix the designs. Crestings are shown in plates 15 and 16. Baluster patterns, as well as individual patterns, were frequently used for fences and gates, which, with grave rails, were common in Adelaide (plates 17–19). Plates 20 and 21 show two other types of ornamental castings illustrated in the catalogues.

PLATE 1—The title page of the catalogue of Stewart and Harley's "Sun" Foundry, Hindley Street, Adelaide, first edition, 1897[12]

PLATE 2—The title page of the second edition of the catalogue of the "Sun" Foundry of A. C. Harley and Co., 1914[13]

"SUN" FOUNDRY.

ILLUSTRATED CATALOGUE.

COLIN STEWART.

FIRST EDITION.

Vol. I.

ALLAN CAMERON HARLEY.

Architectural, Sanitary, and General Castings, and Wrought Ironwork.

STEWART & HARLEY,

General Ironfounders and Blacksmiths.

HEAD OFFICE and WORKS:

"SUN" FOUNDRY, HINDLEY ST., ADELAIDE, SOUTH AUSTRALIA.

ADELAIDE:
Printed at the Crescent Printery, 6 & 8, Topham Street (opposite Harrold, Colton, & Co.)
1897.

"SUN" FOUNDRY.

Illustrated Catalogue.

SECOND EDITION. VOL. II.

ALLAN CAMERON HARLEY, SENR.
PRINCIPAL.

HENDRY RISK HARLEY.
COMMERCIAL.

WILLIAM KIRK HARLEY.
FOUNDRY EXPERT.

ALLAN CAMERON HARLEY, JUNR.
WORKS MANAGER.

Architectural, Sanitary, and General Castings, and Wrought Ironwork.

A. C. HARLEY & CO.,

GENERAL IRONFOUNDERS AND BLACKSMITHS.

HEAD OFFICE AND WORKS:

"SUN" FOUNDRY, HINDLEY STREET, ADELAIDE, SOUTH AUSTRALIA

ADELAIDE:
PRINTED BY VARDON & SONS, LIMITED, GRESHAM STREET.

No. 956.

No. 957.

No. 958.

No. 959.

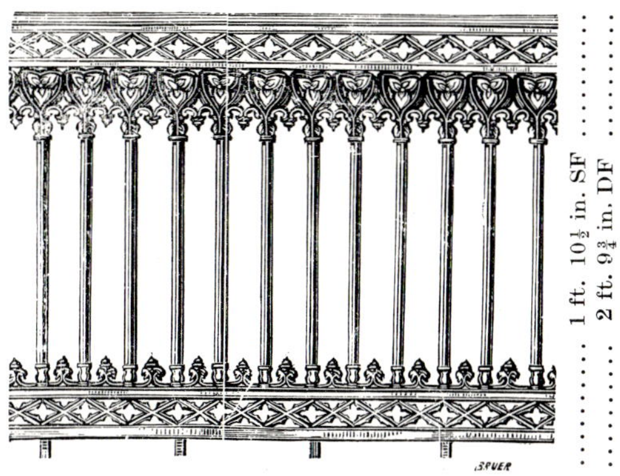

No. 960.

Plates 3–21 reproduced from the second edition of the catalogue of the "Sun" Foundry of A. C. Harley and Co.[13]

PLATE 3—Verandah and Balcony Railings, shown on page 102, Harley's catalogue. An introductory note to this section of the catalogue states: "Verandah and Balcony panels, with and without Balusters, the width of which are shown, are sold at per each panel and each Baluster (which may be spaced to suit purchasers). All others are continuous, and are sold at per foot lineal." The design of No. 957 is based upon circles and stylised acanthus leaves. *(Acanthus spinosus* was favoured by the Greeks, while *Acanthus mollis,* without spines, is said to have been favoured by the Romans.) Still more fragmented are the acanthus leaves shown in No. 956

No. 557.

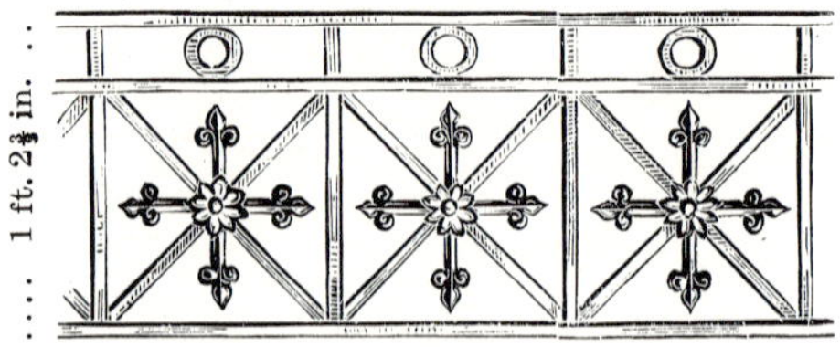

No. 558.

No. 559.
Each Panel 1 ft. 8 in. wide ; each Baluster 5 in. wide.

No. 560.
Each Panel 1 ft. 1⅜ in. wide ; each Baluster 2¾ in. wide.

No. 561.

No. 562.

No. 563.

No. 564.

PLATE 4—Verandah and Balcony Railings (p. 103, H.C.)

No. 961.

2 ft. 10 in.

No. 962.

2 ft. 6¼ in.

No. 963.

2 ft. 11¼ in.

No. 964.

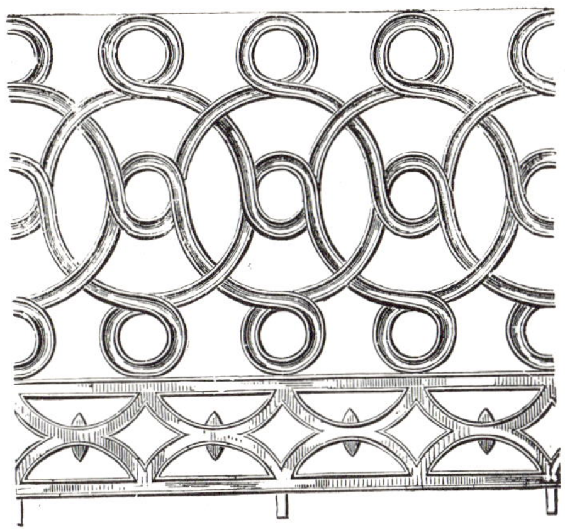

2 ft. 9¾ in.

No. 965.

2 ft. 9½ in.

No. 966.

2 ft. 8½ in.

PLATE 5—Verandah and Balcony Railings (p. 104, H.C.)

No. 565.

No. 566.

No. 567.

No. 568.

No. 569.

No. 570.

PLATE 6—Verandah and Balcony Railings (p. 105, H.C.)

No. 967.

No. 968.

No. 969.

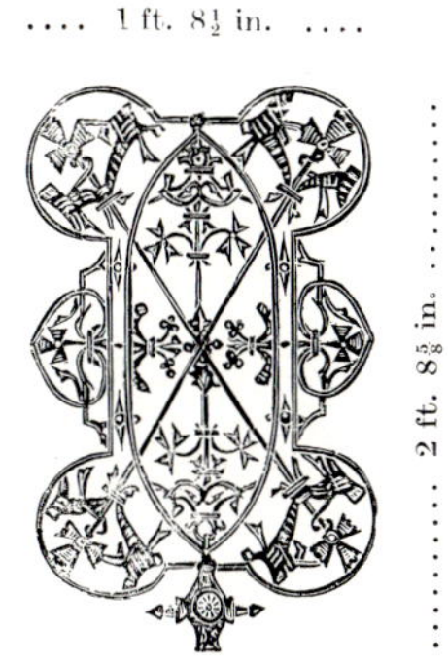

No. 970.

No. 971.

No. 972.

No. 973.

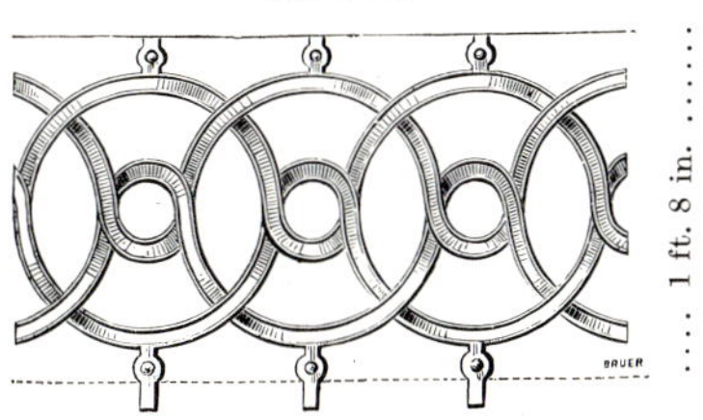

No. 974.

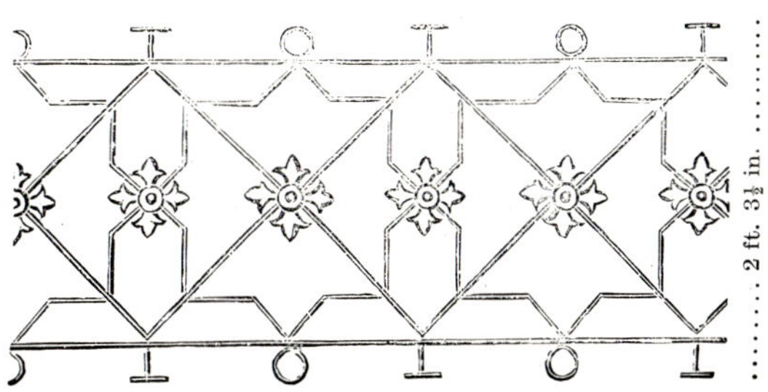

PLATE 7—Verandah and Balcony Railings (p. 106, H.C.)

No. 571.

No. 572.

No. 573.

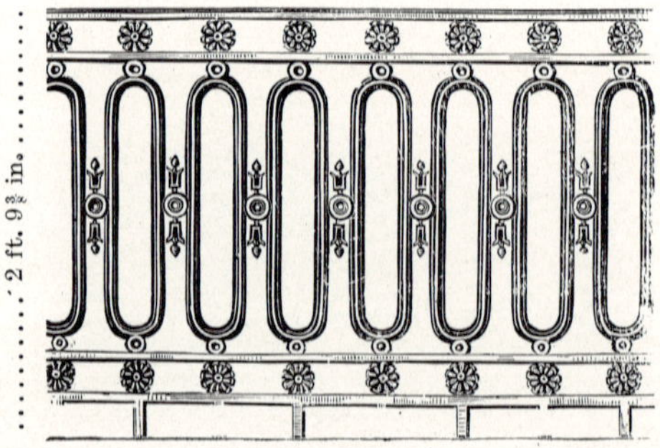

No. 574.

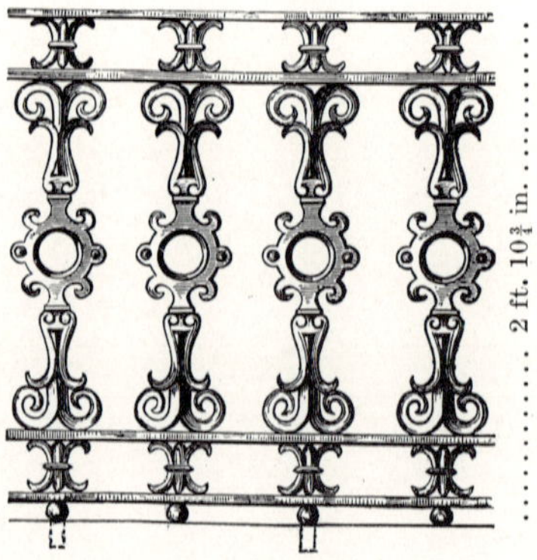

No. 575.

No. 576.

Each Panel 1 ft. 7⅜ in. wide.

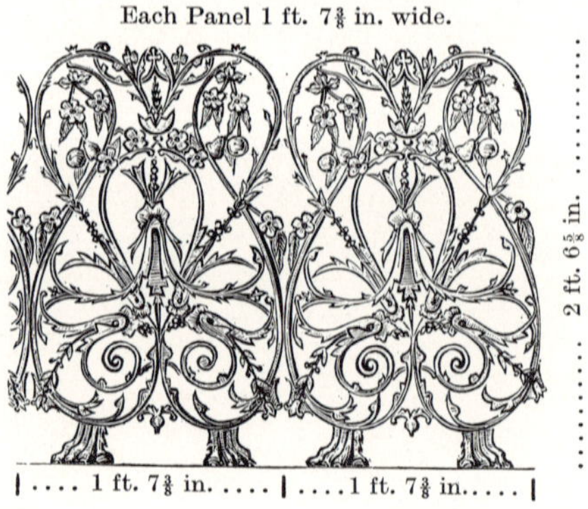

PLATE 8—Verandah and Balcony Railings (p. 107, H.C.). Number 576 shows a mixture of acanthus foliage, flowers, fruit, and claw feet

No. 577.

No. 578.

No. 579.

No. 580.

Each panel 1 ft. 5½ in. wide.

No. 581.

No. 582.

Each Panel 1 ft. 5⅜ in. wide; each Baluster 4½ in. wide.

PLATE 9—Verandah and Balcony Railings (p. 108, H.C.). Number 579 prominently displays the stylised palmette, while No. 580 shows ground-fern fronds growing from a dwarf tree-fern trunk

No. 583.

No. 584.

Each Panel 2 ft. $0\frac{1}{2}$ in. wide ; each Baluster $2\frac{1}{8}$ in. wide.

No. 585.

1 ft. $3\frac{1}{2}$ in.

No. 587.

1 ft. $7\frac{3}{4}$ in.

No. 586.

Each Panel 2 ft. $7\frac{3}{4}$ in. wide ; each Baluster $1\frac{7}{8}$ in. wide.

No. 588.

1 ft.

PLATE 10—Verandah and Balcony Railings (p. 109, H.C.)

No. 589.

No. 590.
Each Panel 1 ft. 4⅝ in. wide ; each Baluster 1½ in. wide.

No. 591
Each Panel 1 ft. 6¼ in. wide ; each Baluster 3 in. wide.

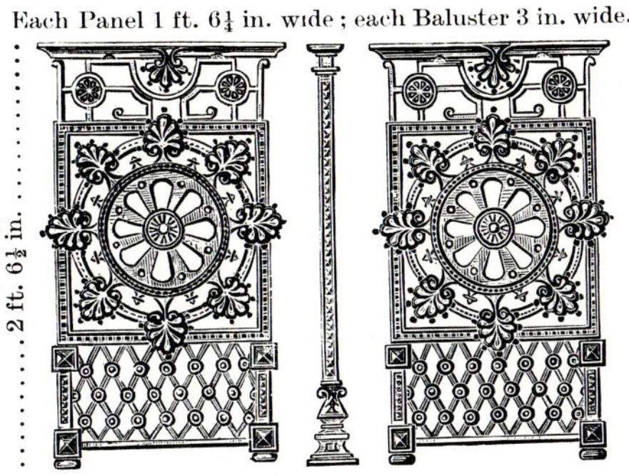

No. 592
Each Panel 1 ft. 6 in. wide ; each Baluster 3 in. wide.

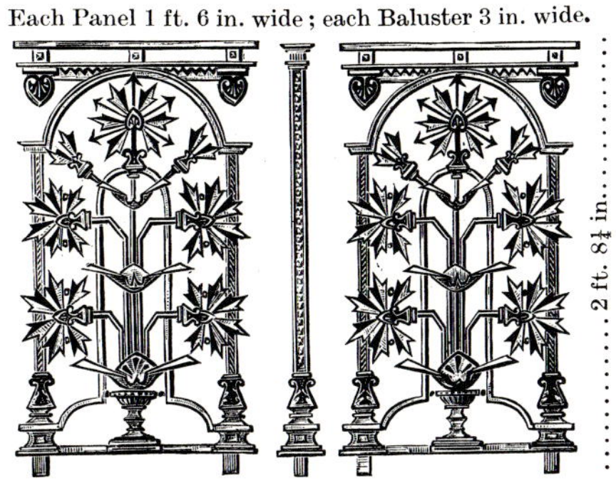

No. 593.
.... 1 ft. 6⅛ in. ....

No. 594.
Each Panel 1 ft. wide ; each Baluster 2⅛ in. wide.

This Panel can be had straight or curved, as Sections Nos. 1 and 2.

No. 1.

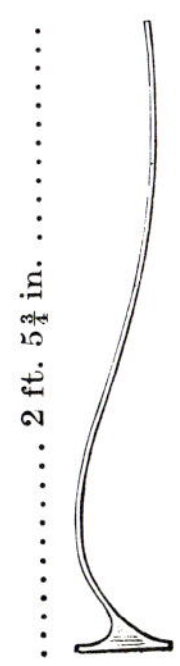

No. 2.

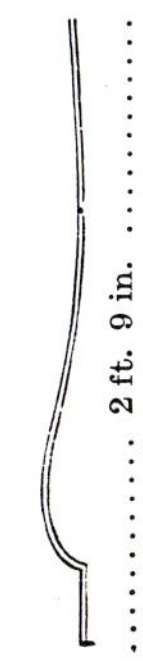

Plate 11—Verandah and Balcony Railings (p. 110, H.C.)

PLATE 12—Brackets and Double Brackets, with Frieze and Fringe combined (p. 91, H.C.). One of the fifteen pages of illustrations of brackets in the "Sun" catalogue, second edition[13]. Daisies are shown in Nos. 422 and 418A, fern fronds in 423, and palmettes in 420A

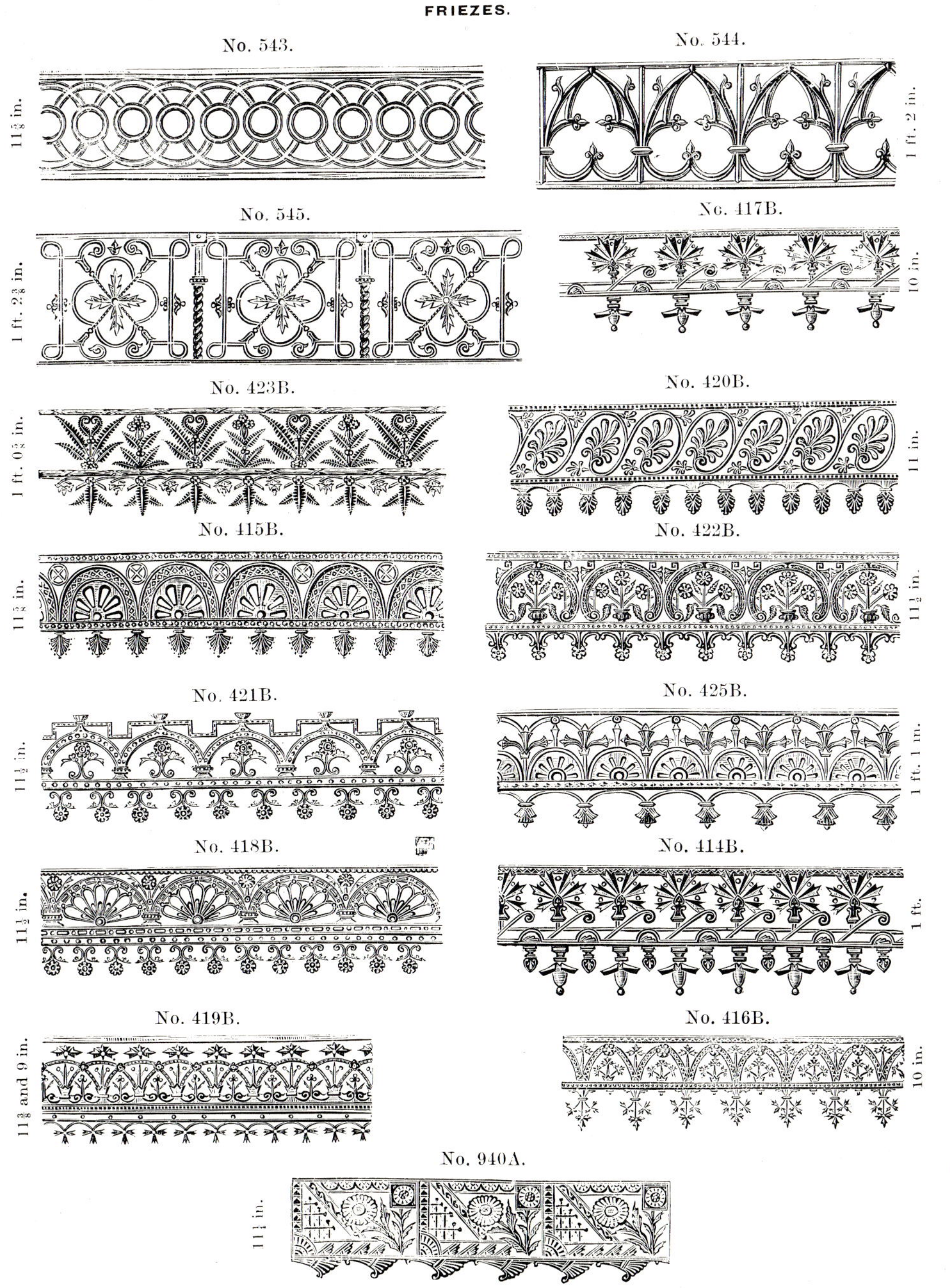

PLATE 13 Friezes (p. 101, H.C.). Number 940A was added in the second edition. One of six pages of illustrations of friezes. Sunflowers are represented in No. 940A

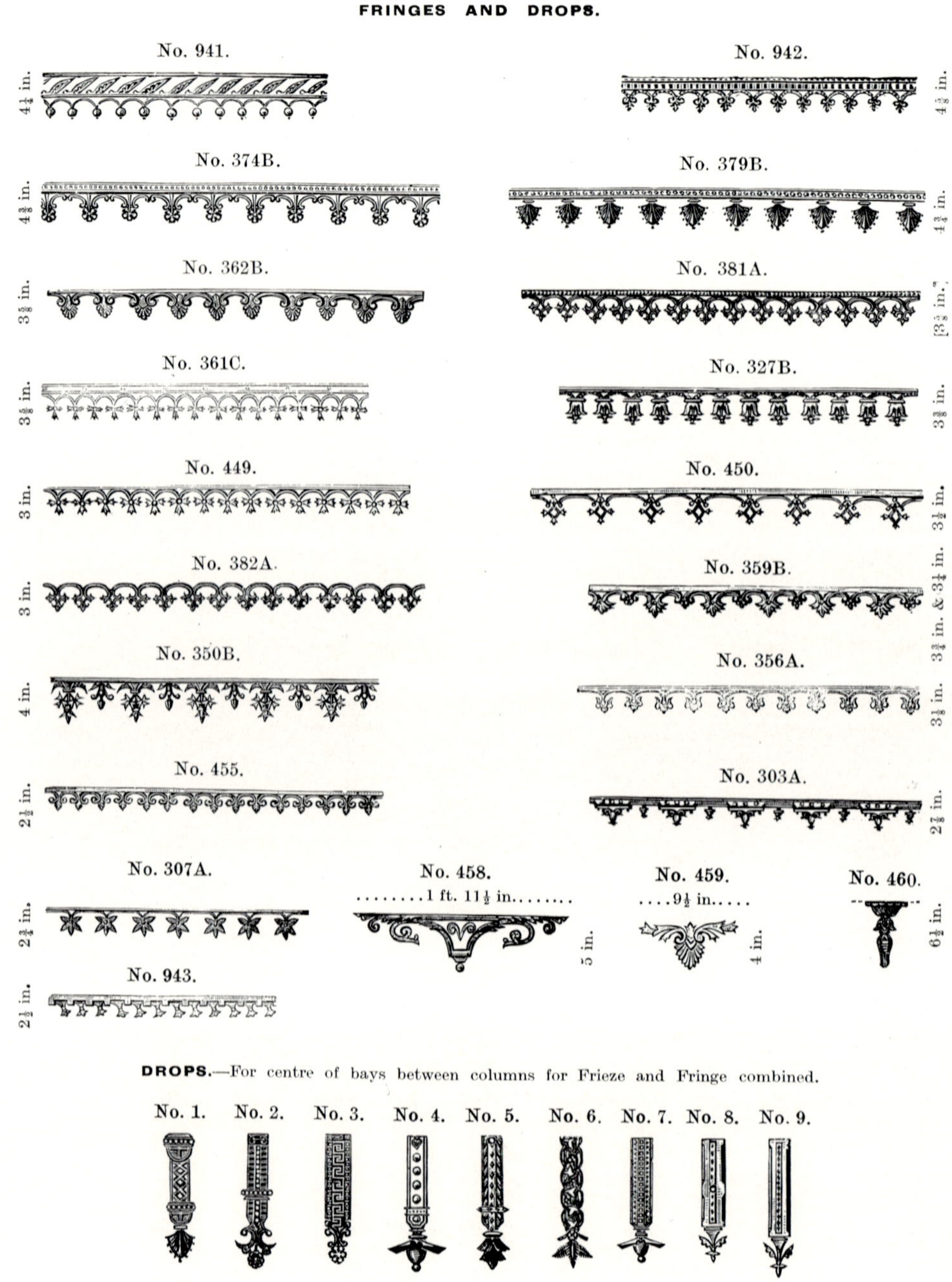

PLATE 14—Fringes and Drops (p. 95, H.C.)

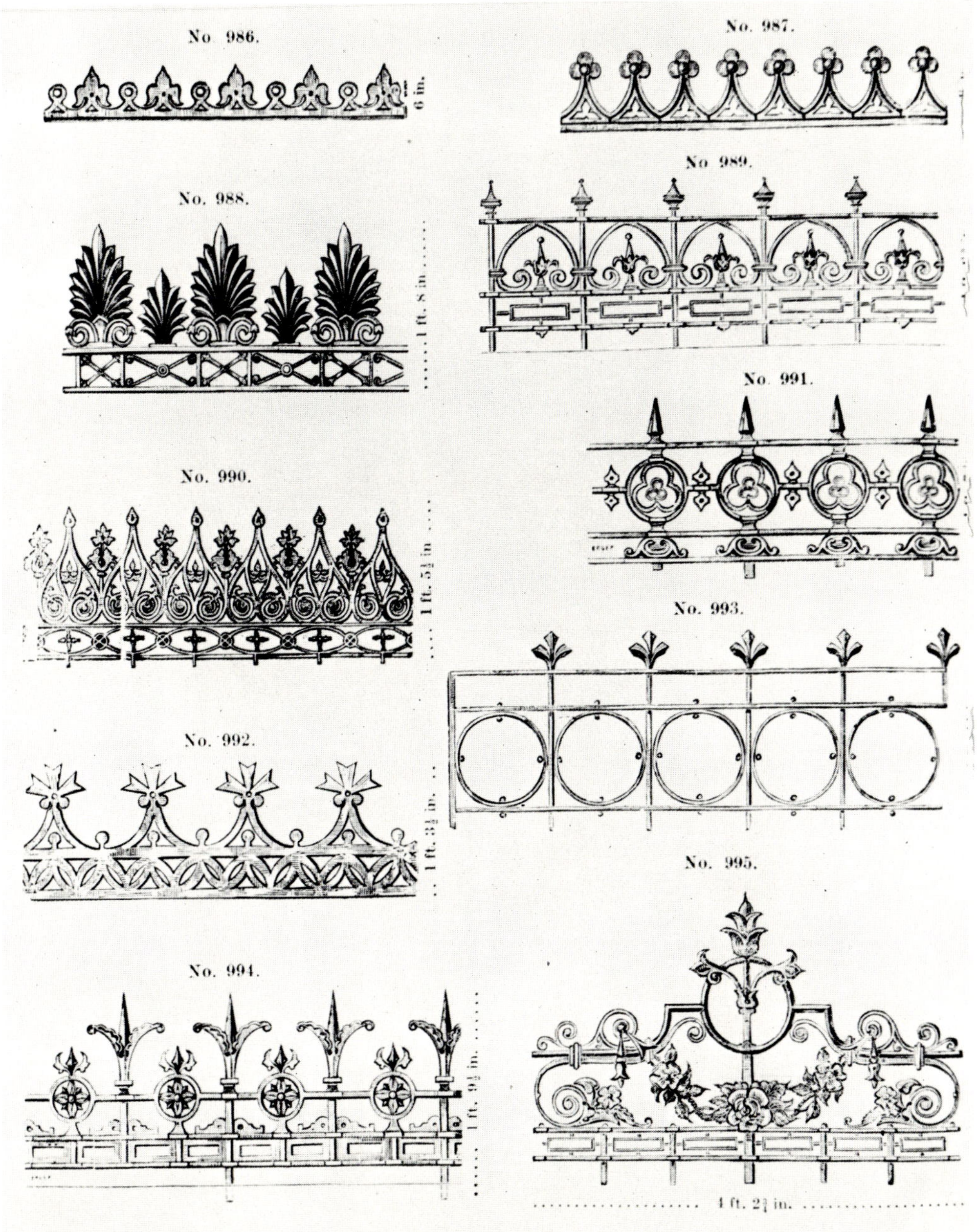

PLATE 15—Crestings (p. 117, H.C.). Palm leaves form the chief element of No. 988, while roses form a swag in No. 995

PLATE 16—Crestings (p. 120, H.C.)

CAST IRON FENCES.

No. 227.

No. 228.

No. 229.

May be prepared to fix with or without Intermediate Standards, from 2 ft. 10 in. centres or wider.

No. 230.

Standards can be set at any distance required.

No. 231.

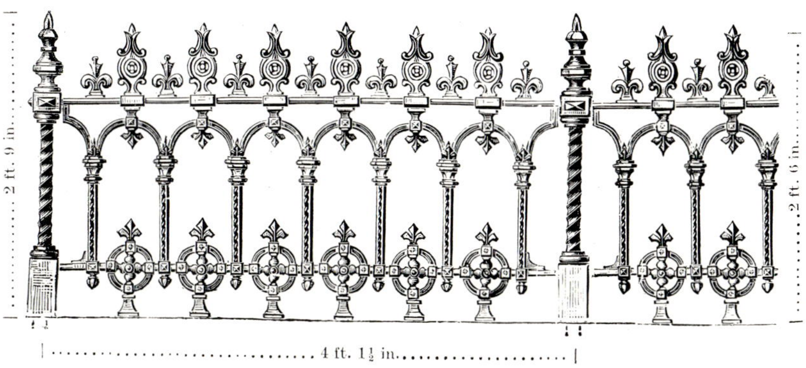

PLATE 17—Cast Iron Fences (p. 47, H.C.). Eight pages of the catalogue are devoted to fences. Some patterns are similar to those illustrated as balcony railings, some being shorter, and others about a foot taller, than the corresponding balusters

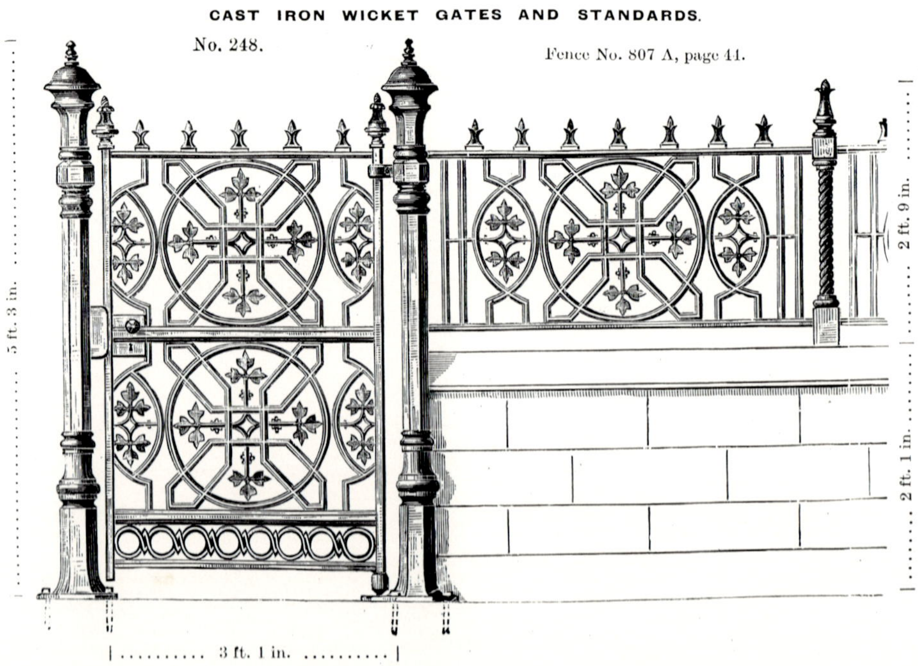

PLATE 18—Cast Iron Wicket Gate and Standards (pp. 56 and 55, H.C.). Cast iron gates and fences are of more varied design than in other cities and indeed are more common than balustrades. The pattern in No. 248 is the same as the grave railing shown in plate 19, while that in No. 227 is also shown in No. 227, plate 17. Twenty-seven pages are devoted to gates, many with matching fences. A few of the gates are of wrought iron and some of these have cast iron caps to the pickets

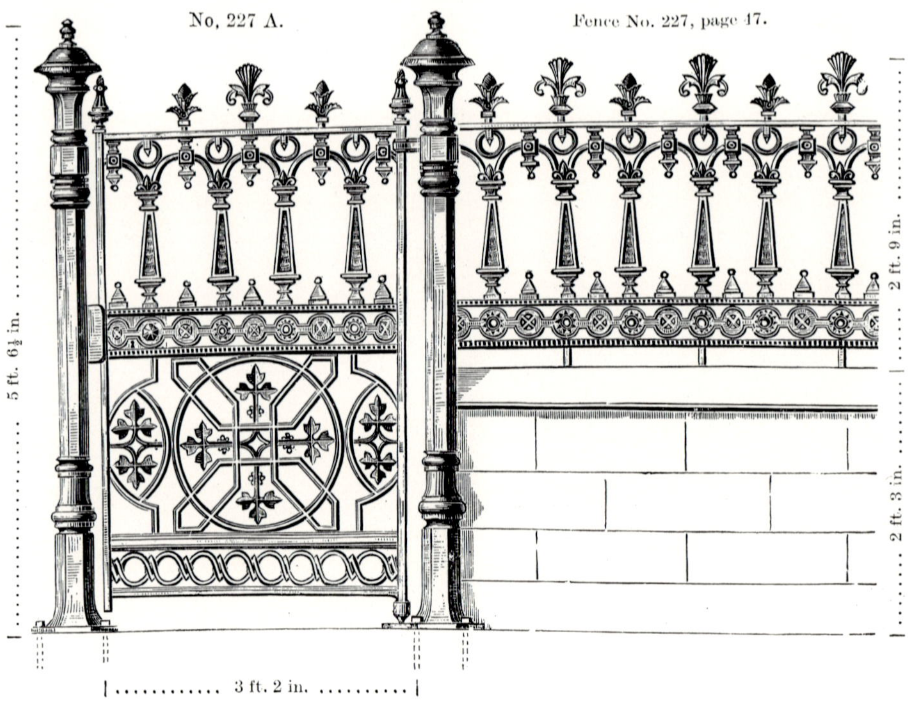

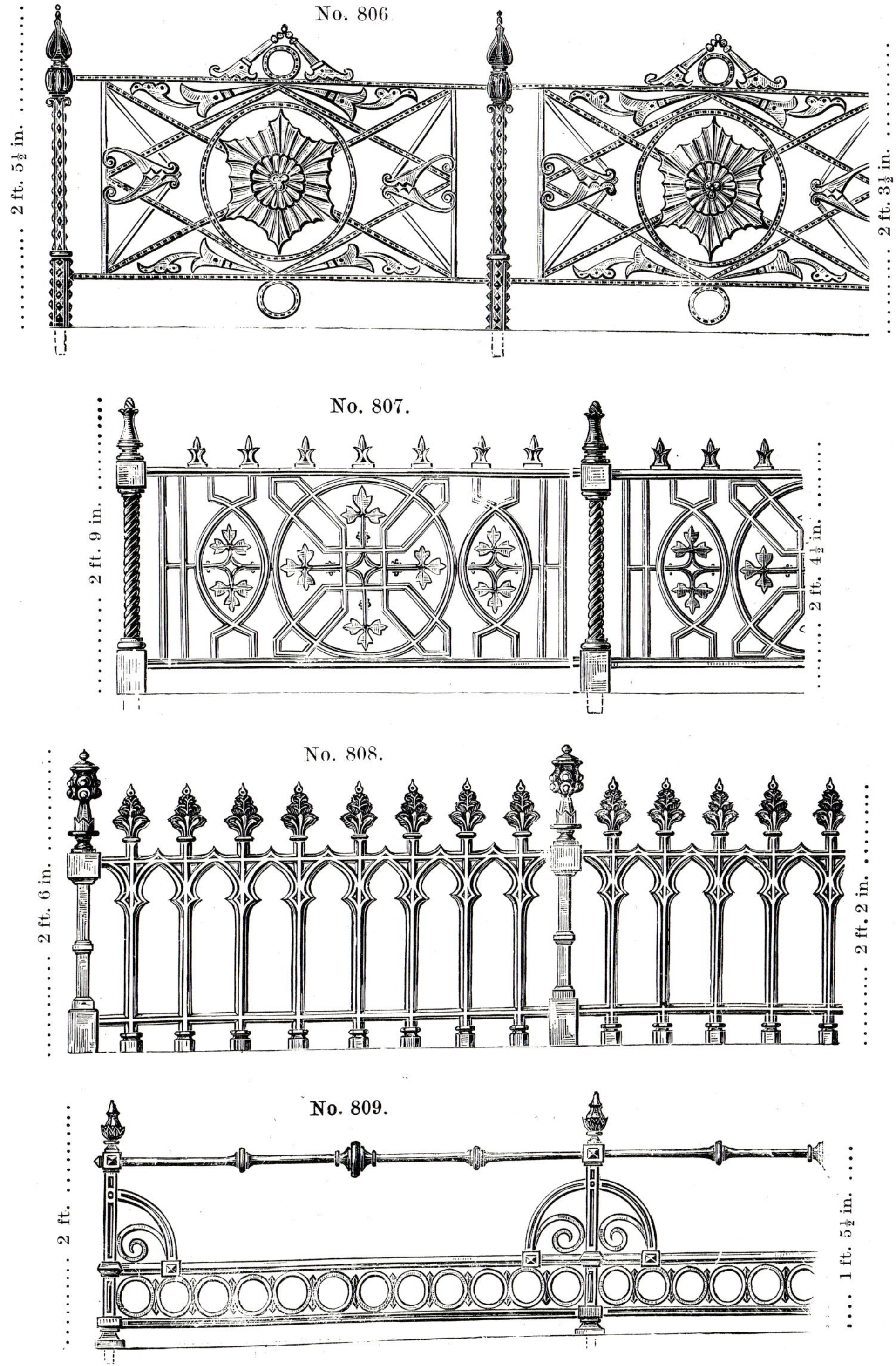

PLATE 19—Grave Fences (p. 136, H.C.). Eleven pages are devoted to grave fences, a usage of cast iron highly characteristic of the period throughout the world. It is said that the "Sun" foundry was responsible for almost all the cast iron fences in Adelaide cemeteries, while the grave of one of its founders, Colin Stewart, had a wooden fence

CRESWELL GARDEN

PLATE 20—An arch for Creswell Garden, and the arch as it exists today (between King William Road, North Adelaide, and the Adelaide Oval)

PLATE 21—Canopy and drinking fountain (p. 5, H.C.). This is also shown in the Fulton catalogue

CATALOGUE OF FULTON'S CASTINGS

G. E. Fulton conducted a foundry at Wilpena Terrace, Kilkenny, for a number of years, and there made castings of many types, while specialising in the vertical castings of large pipes. The firm published at least two catalogues. No copy of the first edition has been sighted. The second edition was published in 1887[14], ten years before Stewart and Harley's "Sun" Foundry catalogue. The title page is shown as plate 22.

* * *

Many of the patterns shown in Fulton's catalogue appear in the later Harley catalogue, sometimes with a slight increase in measurements.

Some examples of the illustrations in Fulton's catalogue are shown in plates 23–26, and in plates 95–99. Plates 23–25A show columns and posts made by the firm. The capitals are based on those of classical antiquity, Tuscan, Doric, Ionic, Corinthian, and Composite. The columns are much more varied than those to be seen in Victoria, although in both states modified Corinthian capitals were most commonly used. Plates 25B, 26A, and 26B show overthrows with lanterns, finials, and weather vanes.

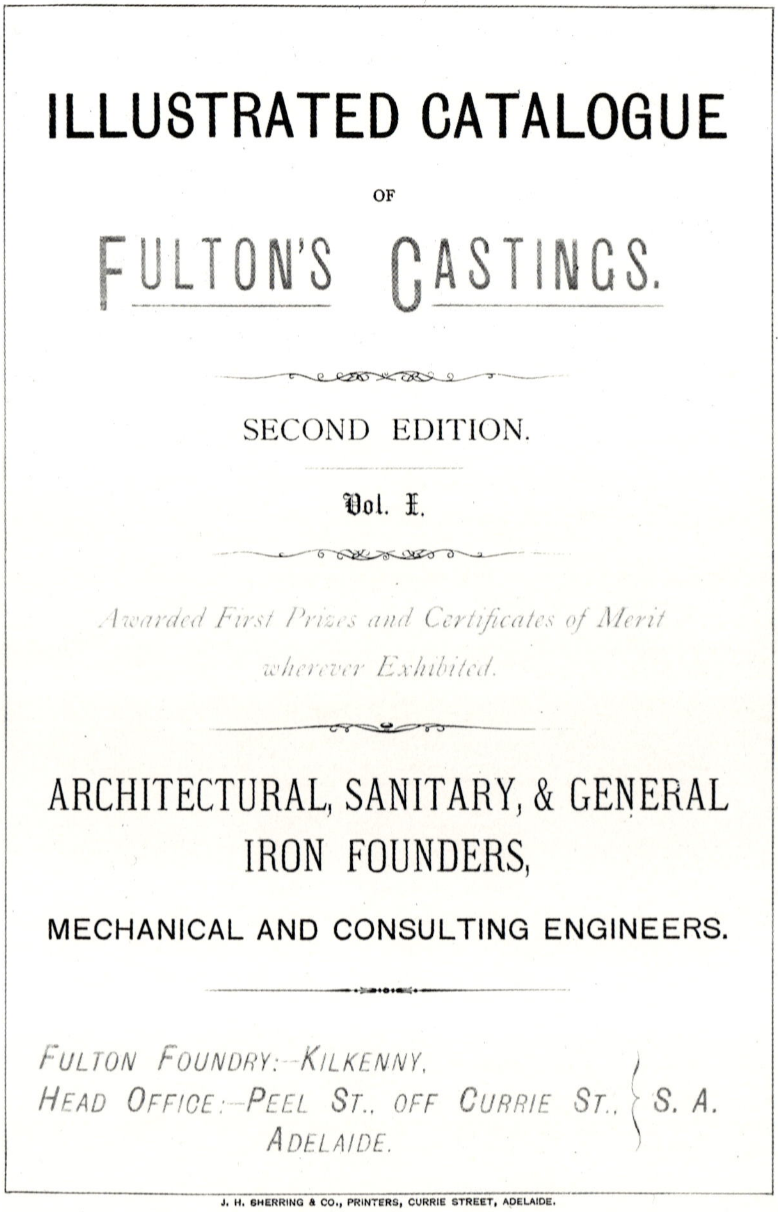

ILLUSTRATED CATALOGUE

OF

FULTON'S CASTINGS.

SECOND EDITION.

Vol. I.

Awarded First Prizes and Certificates of Merit wherever Exhibited.

ARCHITECTURAL, SANITARY, & GENERAL IRON FOUNDERS,

MECHANICAL AND CONSULTING ENGINEERS.

FULTON FOUNDRY:—KILKENNY,
HEAD OFFICE:—PEEL ST., OFF CURRIE ST., ADELAIDE. } S. A.

J. H. SHERRING & CO., PRINTERS, CURRIE STREET, ADELAIDE.

PLATE 22—Title page of the second edition of *Illustrated Catalogue of Fulton's Castings* dated 1887[14]. The Preface states: "We have much pleasure in submitting the Second Edition of our Illustrated Catalogue. A perusal of it will show that we have spared no pains or expense in its production, and we trust that the same success will attend it as accompanied the First Edition.

"Notwithstanding the fact that we have more than doubled the size of the Catalogue, it has been found impossible to include all our patterns, as we are constantly adding to their number.

"We specially invite personal enquiry as to requirements not shown in this book, as we believe that we are in a position to produce any design in Cast-Iron of a quality and at a price that will compete with the very best English makers.

"We make our grateful acknowledgements to the Public, the Architectural Profession, and our Customers generally, for their liberal support and aid in the past."

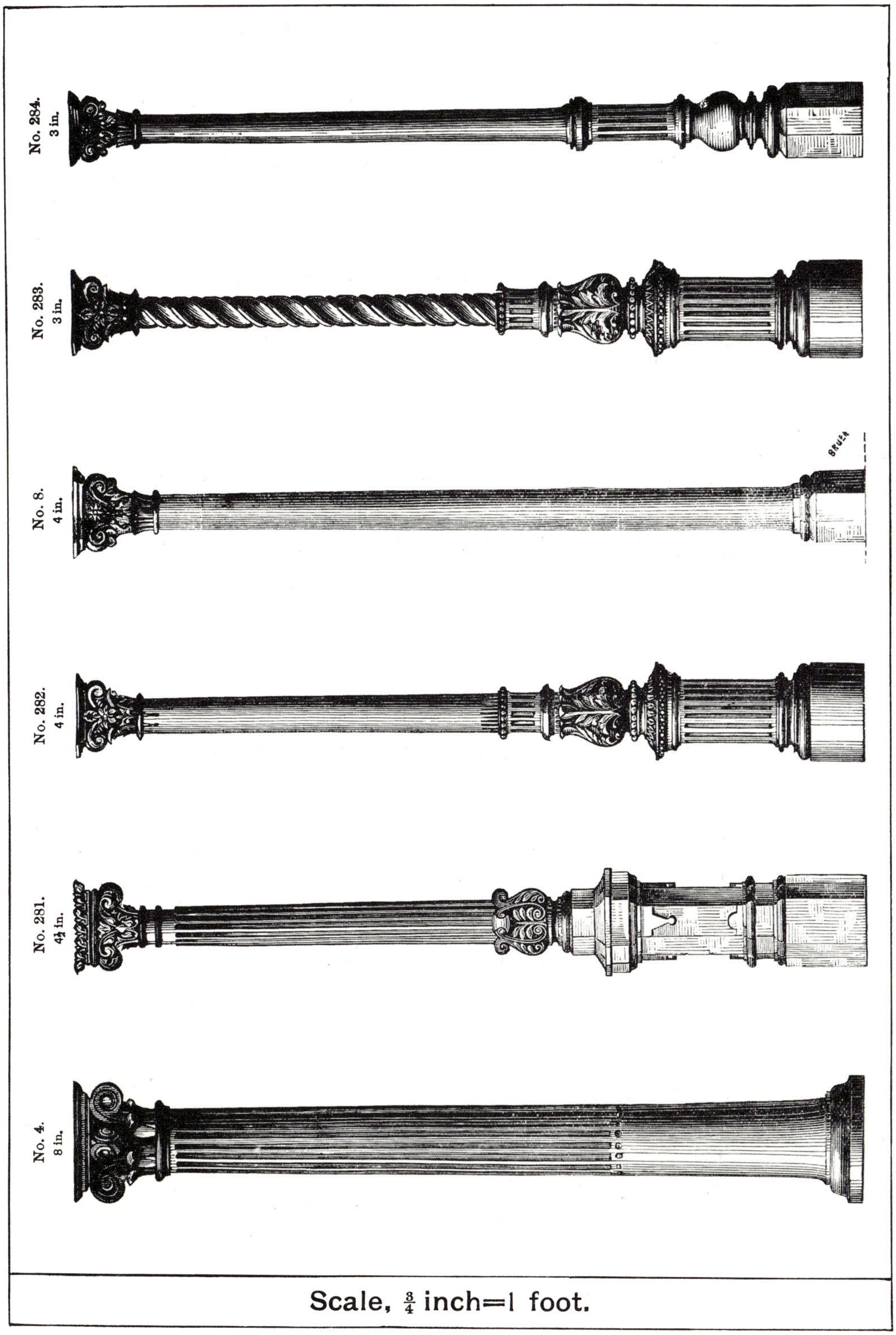

PLATE 23A—Verandah columns, shown on pages 47 and 48 of Fulton's catalogue. The two "drops" for pilasters (Nos. 7 and 8, in 23B) are shown also in Harley's catalogue

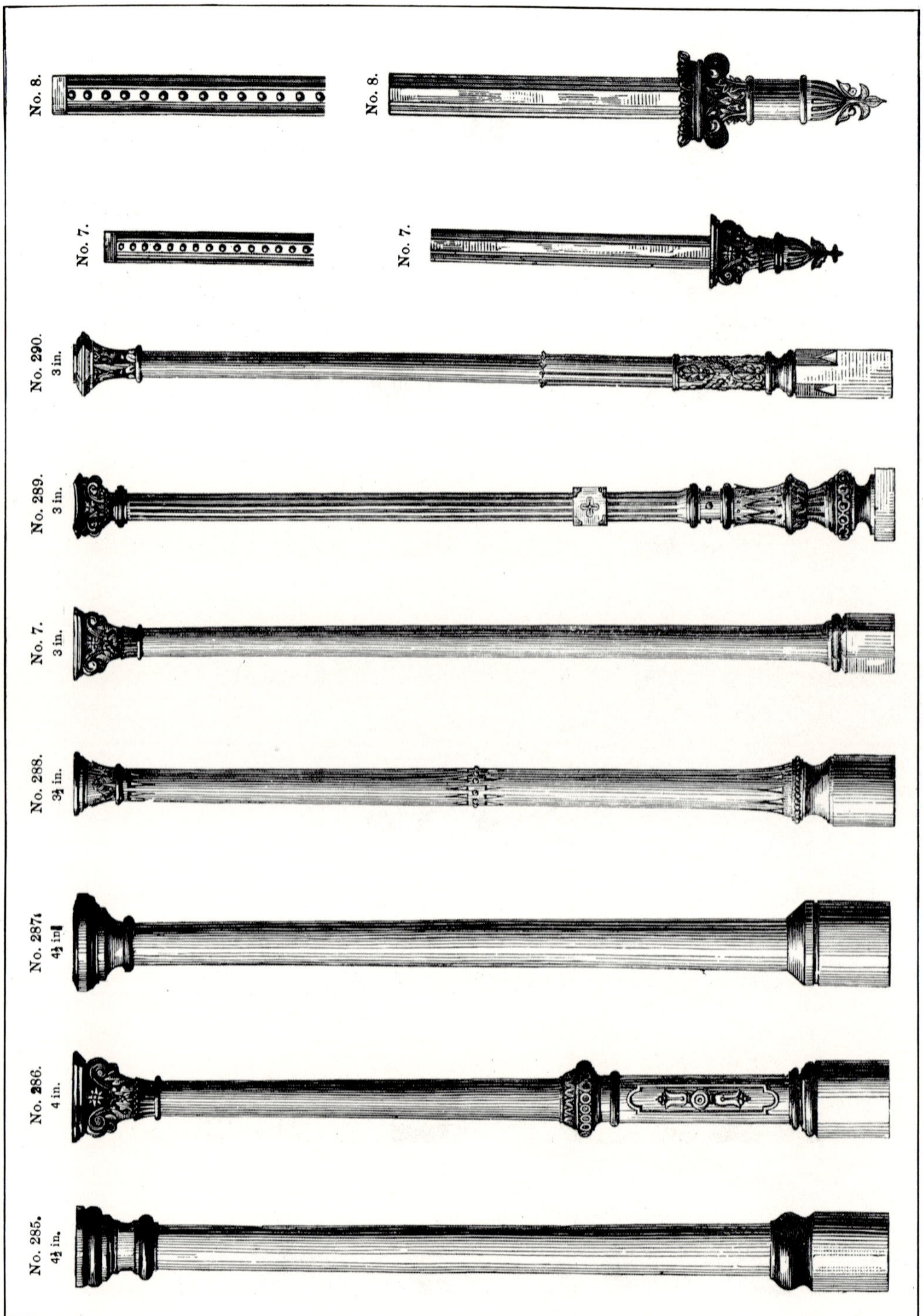

Scale, ¾ inch=1 foot.

PLATE 23B

PLATE 24A—Verandah columns, bridle and newel posts, shown on pages 45 and 49 of Fulton's catalogue. Seven of the seventeen verandah posts shown in plates 23 and 24 are to be found in the thirty designs for columns in Harley's catalogue, while the "Example" shown in plate 24B also appears in both catalogues

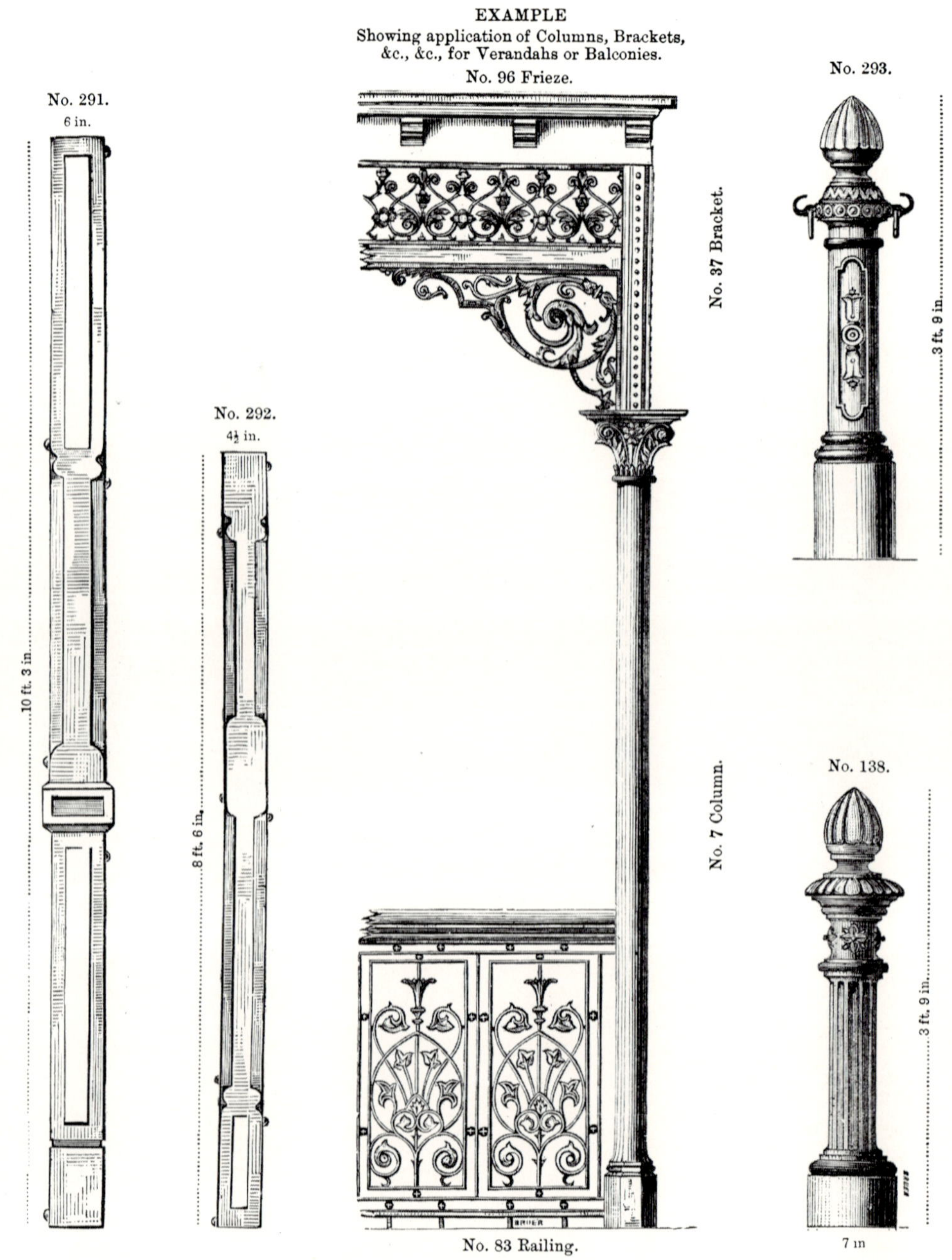

Scale, ¾ inch=1 foot.

Plate 24b

PLATE 25A—The gas-light era (Fulton's catalogue).
Gas-light standards. (The central standard is the same as one in Harley's catalogue, except for the absence of consoles at the base in the latter. The other two are not represented in Harley's catalogue)

PLATE 25B—Gate overthrows and lanterns. (No. 298 is the same as No. 819 in the Harley catalogue, while the arch forms a part of the canopy and drinking fountain shown here as plate 21, from Harley's catalogue)

PLATE 26A—Finials (page 39, Fulton's catalogue. Four of the five finials shown are represented in Harley's catalogue. There are twenty-four finials in Fulton's catalogue and fifty-three in Harley's catalogue)

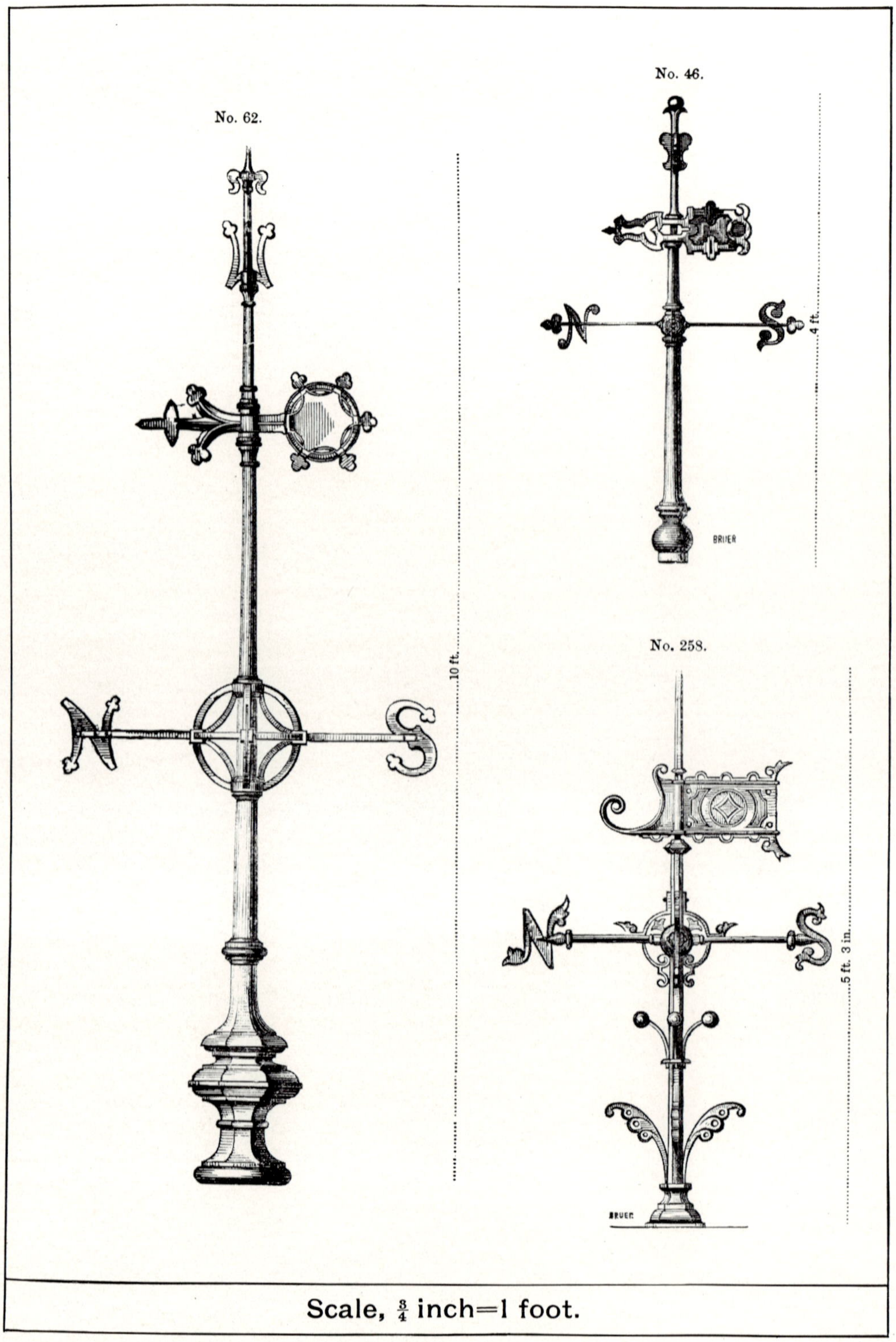

PLATE 26B—Weather vanes (page 40, Fulton's catalogue. This page is merely a rearrangement of the three vanes shown on page 116 of Harley's catalogue, save for the absence of the artist's name, Bruer, from Harley's catalogue, and a difference in the measurements of one vane)

STATE REGISTRATION OF DESIGNS

Another important source of reference is found in the registration of patterns for purposes of copyright. Those registered in Victoria and New South Wales have been reproduced in *Ornamental Cast Iron in Melbourne*[15] and *Sydney Lace*[16]. Most of the South Australian registrations are shown in plate 27A. A few were so poor as to make reproduction unsatisfactory. An unusually clear example is shown in plate 27B. South Australian registrations were few in proportion to other states.

The following is a list of the firms which registered designs in South Australia:

Carr, J. W., 1 design, S.A. No. 35 registered on 1.10.1883—palisading for villas or tombs.

de Mole, W. F. and G. E. Farrar, 1 design, S.A. 23 (29.3.1883)—railing.

Forwood, Down & Co., 1 design, S.A. 10 (7.7.1881)—bracket.

Fulton, G.E. & Co., 8 designs between S.A. 26 (24.10.1883) and 52 (27.2.1885)—railings, gates, bracket, and ventilator cover.

Garlick, D., 8 designs, between S.A. 1 (4.3.1880) and 30 (7.3.1883)—iron railings and gates, brackets.

Gray, W.F., 1 design, S.A. 2 (12.12.1878)—ornamental gutters.

Hill & Pearce, 1 design, S.A. 48 (26.7.1884)—iron palisading and gates.

Mellor Bros., 1 design, S.A. 27 (13.6.1883)—railing head.

Revell, Adams & Co., 2 designs, S.A. 21 (4.7.1883) and 33 (9.9.1883)—balcony railings.

Simpson, A. & Son, 1 design, S.A. 6 (29.4.1880)—window sill ventilator.

Stewart Jones & Co., 4 designs, S.A. 57–60 (26.1.1886)—heads for railing and ornamental ironwork.

Undoubtedly many other firms produced ornamental castings.

THE MODES OF USAGE OF CAST IRON ORNAMENTATION IN ADELAIDE

The modes of combining the component parts of cast iron ornamentation of Adelaide's verandahs and balconies are illustrated in plates 28–36. Examples applied to buildings are shown in plates 37–94. Some of the usage which differs from that in other States receives particular comment. Wherever possible the manufacture of the castings is documented. In some instances the pattern is also to be found in the catalogue of the Saracen Foundry of Walter Macfarlane, of Possilpark, Glasgow.

A tour of inspection may commence at the Botanic Hotel at the corner of North and East Terraces, continue along part of East Terrace and then back along Rundle and Hindley Streets to West Terrace, so to regain North Terrace which is followed back to the Botanic Hotel (plates 42–68). Other buildings in Adelaide, North Adelaide, and further afield, complete this section.

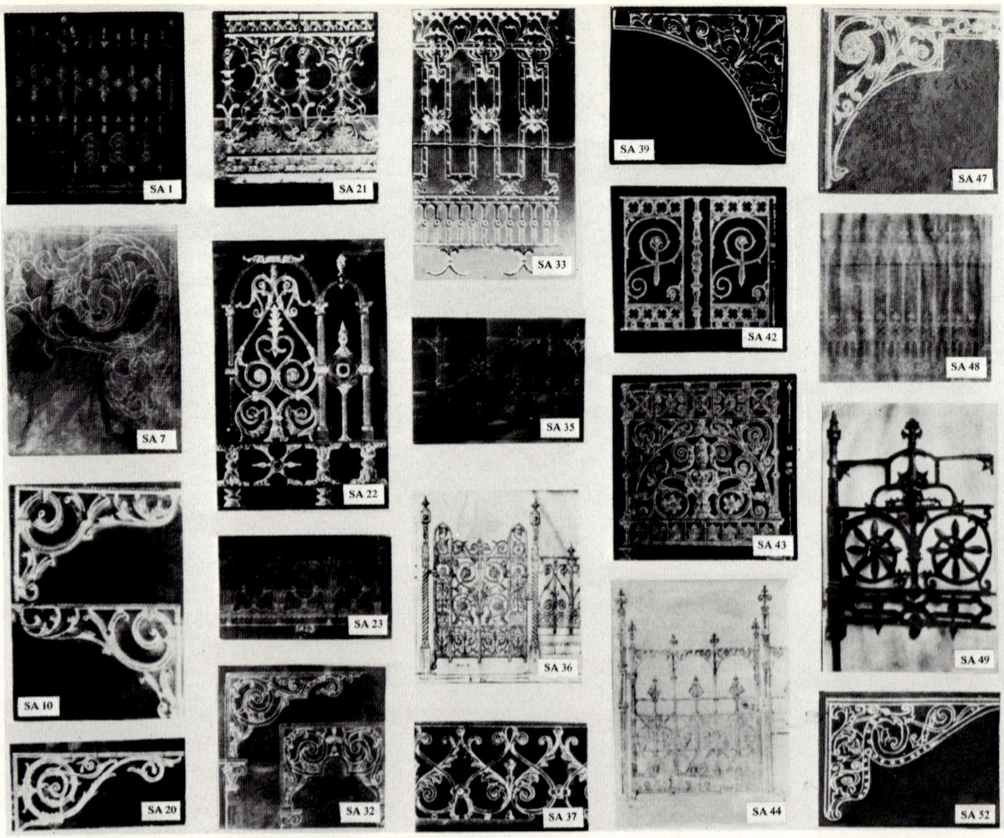

PLATE 27A—South Australian registration of designs "for ornamentation of verandahs, balconies, etc." In a few cases poor originals precluded reproduction. Columns to be read downwards:

First column: No. S.A. 1; D. Garlick (applicant), 4.3.1880. 7; D. Garlick, 17.11.1880. 10; Forwood, Down & Co., 7.7.1881. 20; D. Garlick, 9.10.1882

Second column: 21; Revell, Adams & Co., 4.7.1883. 22; D. Garlick, 21.3.1883. 23; W. F. de Mole & G. E. Farrar, 29.3.1883. 32; D. Garlick, 1.9.1883

Third column: 33; Revell, Adams & Co., 9.9.1883. 35; J. W. Carr, 1.10.1883. 36; G. E. Fulton & Co., 24.10.1883. 37; D. Garlick, 11.12.1883

Fourth column: 39; G. E. Fulton & Co., 2.4.1883. 42; G. E. Fulton & Co., 15.7.1884. 43; G. E. Fulton & Co., 16.7.1884. 44; G. E. Fulton & Co., 16.7.1884.

Fifth column: 47; G. E. Fulton & Co., 16.7.1884. 48; J. Hill & A. G. Pearce, 26.7.1884. 49; D. Garlick & Son, 22.8.1884. 52; G. E. Fulton & Co., 25.6.1887. Ventilator covers, ornamental gutters, window sill ventilator, and railing heads are not shown

PLATE 27B—The mode of registration, showing one of the clearest designs (S.A. 22).
Registrations were fewer in South Australia than in New South Wales, Victoria, and Queensland

Index Heading.	Applicant.	Class.	Number.
Railing.	D. GARLICK		S.A. 22

Date of Application 21 / 3 / 1883

Date of Registration / /

in respect of Cast iron balustrade.

Plates 28–37 show the common components of the ornamental ironwork for verandahs and balconies

Plate 28—Balusters are usually the most noticeable of the decorative elements. They are usually between eighteen and twenty inches wide. In many instances they are set together to make a continuous array, but occasionally they are separated, in which case a narrow intervening baluster improves the appearance

Plate 29—Sometimes columns are set in pairs, necessitating spandrels. On the Newmarket Hotel an unusually wide panel intervenes between two columns. The pattern is similar, but not quite identical, with No. 584 of Harley's catalogue, shown in plate 10 in this book

PLATE 30—Many of the component parts of richer ornamentation are to be seen at 9, 11, and 13 Norwood Parade. The tall brackets are boldly designed botanical volutes which combine to form a comely arch. The deep frieze is framed by wood. Spandrels, of two patterns, are placed between the paired columns. (The balustrade was registered as S.A. 21 by Revell, Adams & Co. on 4.7.1883, and appears in the two editions of the Harley catalogue; No. 565, p. 105, second edition. The capitals are No. 218 in the catalogue. The leaves and stem of the flower between the scrolls may be seen)

PLATE 31—In marked contrast, a minimal use of iron is seen on a row of cottages in Wiel Street, near the Mosque. As is frequent in Adelaide, the columns and capitals are made of wood. The sinuous brackets relieve the severity of the façade. (The pattern is No. 26, F.C.)

PLATE 32—Frieze, bracket, and fringe of 43 Mann Terrace, North Adelaide. A central pediment, with cast iron forming a tympanum, is a frequently used decorative device in Adelaide. (The frieze is No. 488, H.C.; the bracket and matching fringe is No. 360A, and the fringe alone, 360B)

PLATE 33—Deep friezes and brackets of fan patterns combine to ornament a cottage with a portico, having an open pediment with heavy woodwork. If painted, it would be an eye-catching example of over-elaborate decoration. (The bracket, fringe, and frieze are shown separately in the Harley catalogue, and together as No. 418A, p. 91, H.C., shown as plate 12 in this volume. The brackets on the portico are shown as No. 370, p. 86, and the frieze between appears as No. 380B, p. 94. The capitals and column are similar to, but not the same as, those shown in the catalogue)

PLATE 34—Two parallel pediments combine to form a superb piece of Victorian "gingerbread." (The patterns of verandah, balustrade, and fence are somewhat similar. The fence is No. 235, H.C., and the balusters No. 890 H.C., and No. 87, F.C.)

Facing page

PLATE 35—The front of 120 Osmond Terrace, Norwood, with two ornate pediments, the lower profusely ornamented with cast iron. Both have finials

PLATE 36—At the side of 120 Osmond Terrace there is another pediment of somewhat different design, the builder obviously having an affection for variety and elaborate ornament—note the different curves of the roofing. (The triangular pieces at the top and at the two sides are similar to Vic. registration No. 266 and N.S.W. No. 90, except for the addition of curved side pieces)

120

PLATE 37—19 Walkerville Terrace, Medindie. The small triangular pediment lacks function, but tradition is maintained by relating it to the position of the doorway, albeit across the width of the balcony.

Facing page

PLATE 38A—The recessing of balconies is an individual characteristic of Adelaide. The Australian Workers' Union Building (established 1887) shows such an arrangement in its simplest form. The baluster is illustrated in Harley's catalogue (No. 592, p. 110). Iron cresting gracefully caps the tower of the General Post Office in Victoria Square, built between 1867 and 1872, with some later extensions

PLATE 38B—The principle of recession was set early in South Australia, as in the Criterion Hotel, Gawler, where a wooden balcony was recessed. (Photographic print included in *The Gawler Handbook* by G. E. Loyau, 1880)

A

B

PLATE 39—The recession of the balcony roof alone is demonstrated by the Hotel Grenfell, Hindmarsh Square. The tall verandah posts are based on the edge of the footpath. The roof of the verandah, edged by the cast iron balustrade, forms the floor of the balcony. The supporting columns of the roof, and the roof itself, are set well back from the balustrade. This arrangement creates a greater area of balcony exposed to the sun, and the rooms within are lighter

PLATE 40—Detail to show the arrangement of the balcony, Hotel Grenfell. The depth of the verandah was determined by the unusual width of the footpaths of Adelaide, while a narrow balcony was more economical to construct and less weighty to support. (The baluster pattern was registered by Angus Maclean, registered No. Vic. 172, date of registration 15.9.1876. The pattern was also shown in H.C., No. 561, plate 4 in this book, and No. 66, F.C. The verandah column, shown in plate 39, is also in F.C., No. 279, plate 24A)

PLATE 41—Another variant of the stepped-back balcony. A wide bay window, at one time symmetrical around a central door, juts into a narrow unroofed terrace, bounded by very wide balusters, with narrow miniature reproductions with the long axis vertical between the posts

PLATE 42—The Botanic Hotel stands at the corner of North and East Terraces. The dark stone contrasts attractively with the stucco mouldings. The very wide verandah has wooden posts on the edge of the sidewalk. Above, recessed upon the sloping roof of the verandah, a wide balcony runs along two sides of the second storey. The third storey balcony is still further recessed. The hotel was designed by Michael McMullen for R. Vaughan, and built by J. Barry in 1876–1877[17]

PLATE 43—Profile of the Botanic Hotel showing stages of the progressive recession

PLATE 44—East End and East End Market Hotel, East Terrace. The litter in the street indicates its proximity to the fruit and vegetable market, which explains the name of the hotel. The long runs of white-painted balustrades are attractive

PLATE 45—Detail of balustrade of East End Market Hotel (pattern No. 570, p. 105, H.C.)

PLATE 46—Stag Hotel, corner of East Terrace and Rundle Street. It will be noticed that the balcony facing East Terrace is stepped back, while that along Rundle Street is not. (The balcony baluster is No. 578, and that of the tower, No. 226, H.C.)

PLATE 47—Detail of the Stag Hotel showing the cresting on the tower, and an unusual pediment fronting an attic room

PLATE 48—235 Rundle Street. Here a shop balcony is recessed and subdivided to give privacy—a relic of times when living above the shop was common. (The baluster is No. 562, p. 103, and Example p. 96, H.C.; also in F.C., No. 118, p. 6, and Example, p. 49, plate 24B in this volume)

PLATE 49—Austral Hotel, 205 Rundle Street. Along the narrow footpath of Bent Street the balcony accurately tops the verandah, while in Rundle Street, it is recessed. A frieze of coloured glass bears the name of the hotel. The flower on the cast iron frieze resembles the passion-flower, but the number of petals is wrong. (Frieze, No. 528, p. 99, H.C. Brackets, of similar patterns, with minor changes in detail, are shown in H.C., No. 334, p. 82; balustrade, No. 573, p. 107, H.C., plate 6, also No. 88, p. 14, F.C.)

PLATE 50—Hindley Street, Central Adelaide, c. 1879[18]

PLATE 51—The Harrold Brothers' building of plate 50 has been replaced. The Saints building beyond preserves its grace. The ornamental ironwork in the foreground remains. The shop of Coombs, the saddler, houses different wares, while the nearest building remains a hotel, illustrating the contrast between ephemeral and permanent trades

PLATE 52—Tattersalls Hotel, 17 Hindley Street. The building in the foreground of plate 50 is now Tattersalls Hotel, shown also in plate 51. At the beginning of this century the graceful parapet of the earlier era was replaced, and a functionless tower added at considerable expense. The balcony was also recessed in typical Adelaide fashion

PLATE 53—Eagle Hotel, 44 Hindley Street, probably built in 1867 on the site of the Eagle Tavern[19]. A central pediment and tympanum in Adelaide style is incorporated in the recessed balcony. (Frieze No. 528, p. 99, H.C.; matching brackets, No. 338, p. 83; brackets of pediment, No. 348, p. 84; balustrade, No. 571, p. 107, plate 8)

Facing page

PLATE 54—The wide balcony, with a brave display of iron below, of 103 Hindley Street. (Baluster design No. 578, p. 108, H.C., plate 9, a design common in other States. The frieze is No. 532, p. 100, doubled in depth on this store)

JARMANS SHOES

PLATE 55—Hotel Kalgoorlie, 111 Hindley Street, a roofed cantilever balcony is slightly wider than the narrow footpath

PLATE 56—Detail Hotel Kalgoorlie (baluster No. 573, p. 8, plate 8; frieze No. 527, p. 99, and bracket, No. 343, p. 83, all H.C. The baluster is also No. 88, p. 14 in F.C.)

PLATE 57—Kintore Terrace, near the western end of Hindley Street, has a long narrow unroofed cantilever balcony, a type common in London and Sydney, but rare in Melbourne. The word "terrace" may cause some confusion in Adelaide, for it may mean a number of contiguous houses of the same design, or a roadway. The former is the most common usage of the word; the latter is an extension of the definition provided in the *Shorter Oxford English Dictionary* (3rd Ed. revised): "Terrace . . . 1. A raised level place for walking, with a vertical or sloping front or sides faced with masonry, turf, or the like, and sometimes having a balustrade; *esp.* a raised walk in a garden, or a level surface formed in front of a house on naturally sloping ground, or on the bank of a river . . ."

Other usages, such as a gallery, colonnade, portico or flat roof of a house are given, while the fifth definition is relevant to a common use in London and Australia: "5. . . . a row of houses of uniform style on a site slightly, if at all, raised above the level of a roadway." Abutting houses with a common windowless wall between adjacent units are usually implied in architectural usage

PLATE 58—Detail of the balcony of Kintore Terrace with its cast iron consoles. (The baluster pattern is No. 972, p. 106, H.C., reproduced in plate 7 in this book)

PLATE 59—The Newmarket Hotel. A lofty cantilever balcony runs along the West and North Terrace frontages and the chamfered corner between. Some of the decorative fabric is attractive, but some is heavy and clumsy. The two pediments look isolated and give the impression of an unsuccessful attempt at ornamentation for ornament's sake.

Early Adelaide Architecture, by E. J. R. Morgan and S. H. Gilbert, provides the following information[20]: "At this corner on 11 January 1837 Colonel Light began the survey of Adelaide. Opposite was an early market and very soon the New Market Inn was established; by 1854 it was a brick and stone building of ten rooms. The present fine building was put up for Simms and Co. in 1884 to the design of Rowland Rees, architect. Its name was changed to Newmarket Hotel to match its grandeur. The cast iron balustrading to the balcony of the top floor was removed about 1955."

PLATE 60—The cast iron of the balcony of the Newmarket Hotel. The capital is of a common design. (The balustrade was registered as S.A. No. 22, by D. Garlick in 1883—see plate 27 in this book)

PLATE 61—Variations on a theme. The old Morphett Street bridge over the Adelaide railway yards: the way up. The bridge was demolished after this photograph was taken.

PLATE 62—The summit. Holy Trinity Church, North Terrace, is seen above the balustrade. Some wrought iron panels from the recently demolished bridge have been incorporated into the fence of Holy Trinity Church

A

Plates 63a and 63b—The castings were probably made for the South Australian Railways for this specific purpose. The ever-changing shadows of cast iron ornamentation are always fascinating

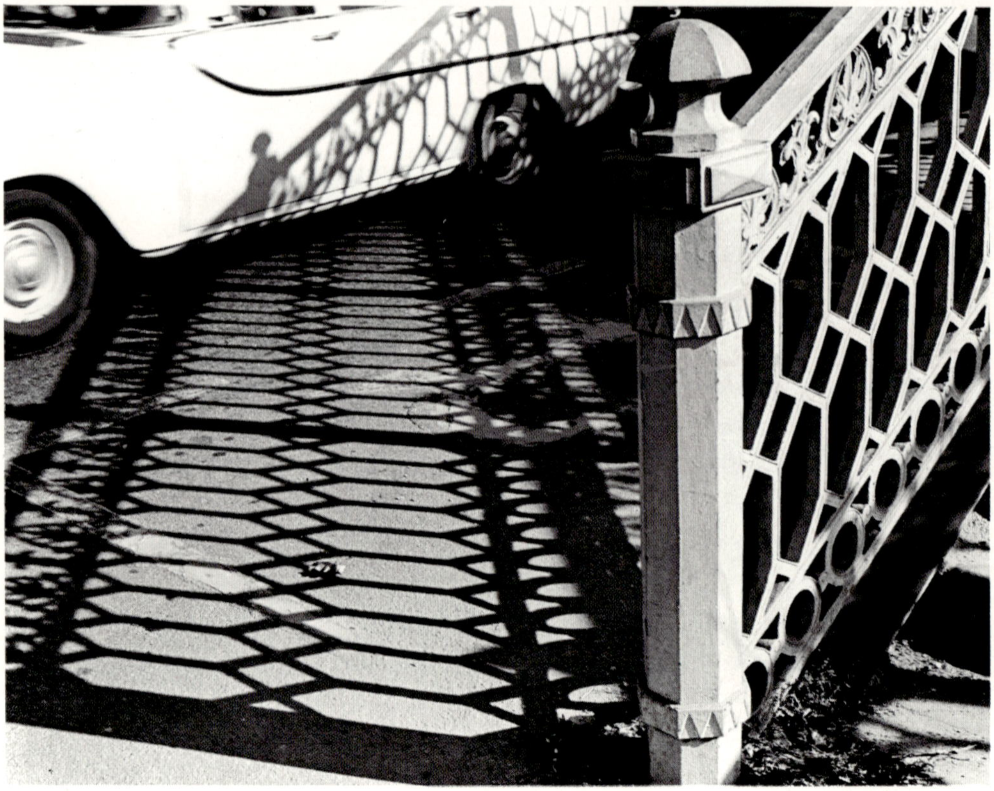

B

PLATE 64—The wide balcony of the Gresham Hotel, North Terrace. Although very wide it was recessed upon the still wider verandah. The transparent roof, the partial enclosure with glass, and the northerly aspect made it a very pleasant place in winter. The demolition of this graceful hotel was a loss to Adelaide

PLATE 65—Cresting above the mansard roof of the Jervois wing of the State Library of South Australia. The foundation stone was laid on 7 November 1879

PLATE 66—Numbers 192, 194, and 196 North Terrace, since demolished. Abutting houses, probably built at the same time, with ornamental stucco work, and ornamented "aprons" paralleling the façades. In one house, protrusions of balcony and verandah fronted bay windows. Four patterns of balusters line verandahs and balconies of the two houses included in the photograph, while the cast iron fence gives unity to a terrace of three houses of varying design. (The baluster of the nearer house, No. 194, is No. 576, p. 107, H.C., plate 8 in this book, which was also registered in Victoria by Holland and Hutchison. Fence, No. 876, p. 42, H.C., and No. 220, p. 17, F.C.)

PLATE 67—The mansard roof with cresting on 261 North Terrace. The cresting is shown as a cast iron fence in Harley's catalogue (No. 884, p. 44). The lower part of the house is shown in plate 111

PLATE 68—219 North Terrace. An attractive house of dark stone with stucco trim. The bay window is capped by a balcony with iron columns, brackets, and friezes, and a wooden balustrade which continues along the main balcony, on which wooden columns are used. The fence supplements the composition (No. 270, p. 75, H.C.)

A

B

C

PLATE 69—Cantilever balconies appeared early in Adelaide architecture. The balustrade shown in these sketches probably was made of wood

(A) King William Street, looking south

(B) King William Street, looking north

from sketches by O. Korn, 1849[21]

(C) Built in 1864, a narrow balcony supported by brackets, with a cast iron balustrade, added to the attractiveness of the Adelaide Club, North Terrace[22]. Most of the balcony has since been removed and the portico stuccoed

PLATE 70 —The balcony of the British Hotel, Finniss Street, North Adelaide, with its supporting cast iron brackets. The façade is striking, with its white painted brick trim contrasting strongly with the stone fabric. Two central long windows give access to the balcony

PLATE 71 —The wide cantilever balcony of the romantically named Seven Stars Hotel, 187 Angas Street. One part bears the date 1855, the other 1894, when the present balcony may have been added

PLATE 72 —A roofed cantilever balcony was sometimes supported by decorative consoles, as on Queen's Chambers, 19 Pirie Street. *Early Adelaide Architecture* states that the building was erected as an office building in 1869. The architects were Garlick and McMinn, and the builders Crocker and Lawson[23]

PLATE 73—Balconettes at 82 to 86 Franklin Street, Adelaide (corner Young Street). The iron is added to a blind window as a purely decorative element

PLATE 74—Number 7 Prescott Terrace, Rose Park, a house of golden stone, with stucco trim. The verandah is unusual in having a rounded gazebo-like protrusion, the only function of which is ornamental. It certainly differentiates this house from all others

PLATE 75—Detail, Prescott Terrace, Rose Park. (The bracket, frieze, and fringe are shown together as No. 422, p. 91, H.C. The drop in the centre of the bay between the columns is No. 3 of a series of nine drops illustrated on p. 95 of the catalogue, shown in this book as plate 14. Such drops are not common in Adelaide. The column is a slight modification of No. 217, p. 40)

PLATE 76—The balcony of private houses may be set back on the verandah, as at 136 Jeffcott Street, North Adelaide. (All patterns are represented in the Harley catalogue: fringe of balcony, No. 350B, p. 95; fringe of verandah, No. 360B, p. 94; upper bracket, No. 308, p. 81; lower bracket cast with fringe, No. 360, p. 84; balustrade, No. 971, p. 106, shown in plate 7. This design was also registered in Victoria, No. 158, on 30.12.1875 by Phillips, McWalter and Chambers. The fence is shown in the Harley catalogue as a grave fence, No. 802, p. 135)

PLATE 77—A fine example of a large house is 97 Jeffcott Street, at the corner of Ward and Jeffcott Streets. It is now the Milford House Private Hospital. The balcony is recessed upon the slightly concave verandah roof. Such a roof would unduly darken the front rooms but, if facing west, would give useful protection from the late sun on a hot summer's day. (The balustrade is No. 567, p. 105, H.C., plate 6; the bracket, No. 298, p. 79, H.C.)

PLATE 78—Park Lodge, corner of Hackney Road and Cambridge Street, Hackney. The circular iron staircase, probably the most economical and compact stairway ever invented, is a recent addition. (Balusters, No. 575, p. 107, H.C., shown in plate 8; frieze, No. 951, p. 96; brackets, No. 336A, p. 87; fence, No. 244A, p. 64, without the geometrical lower one-third of the latter)

PLATE 79—A corner of Admaston, 219 Stanley Street, North Adelaide, built on a steep slope, with a glorious view across tree-tops to the distant hills

PLATE 80—The surprising rear view of Admaston. The present entry floor, shown in plate 79, corresponds with the first floor. (The baluster pattern was registered in New South Wales, No. 89, on 16.8.1881 by Fletcher, Bennett and Frew[24]. The frieze is No. 488, p. 97, H.C., and the bracket and fringe, No. 359, p. 84)

Plate 81—The continuous balcony of 68 Lefevre Terrace, North Adelaide, extends forwards to front the bay as well as the rest of the house (baluster No. 567, p. 105, H.C.)

A

PLATE 82—Pairs of very substantial houses, with cast iron ornamentation, are not infrequent in Adelaide
(A) 58 and 59 Lefevre Terrace, North Adelaide (fence No. 248, p. 56, H.C.)
(B) 64 and 66 Lefevre Terrace, North Adelaide (fence No. 244, p. 53, H.C.)

B

PLATE 83—A row of three houses, 9, 11, and 13 Norwood Parade, of terrace type built to resemble a large single house, has a brave display of ironwork (see plate 30 for details)

PLATE 84—Another terrace of three substantial houses, with verandahs but no balconies, is seen at 74–78 Archer Street, North Adelaide (fence No. 227, p. 47, H.C.)

PLATE 85—Böhm Terrace, 62–68 Archer Street, North Adelaide, built of stone in 1846. Long two-storey ornamented terraces, of the type so common in Melbourne, are rare in Adelaide. The heavy large Victorian bay windows are an unusual feature, and sub-division of verandahs and balconies is a partial one, so that the upper parts of the façade can be seen in a side view. (Baluster No. 561, p. 103, H.C., No. 66, p. 14, F.C.)

PLATE 86—The ironwork of Aquinas College, University of Adelaide (1 Palmer Place) is of unusual type, being mounted on a classical balustrade. *Early Adelaide Architecture* states that[25]: "The central stuccoed part is what remains of 'Montefiore,' the house which was owned and occupied from 1872 to 1916 by Sir Samuel Way, Bt, Chief Justice of South Australia. He bought it from L. M. Cullen and greatly enlarged the existing house, employing William McMinn as his architect. The present entrance front and steps were built about 1880 to McMinn's design. The roof dates from 1916, when an upper storey was removed."

PLATE 87—The ironwork of the verandah of Aquinas College on Montefiore Hill frames vistas of Adelaide. How much better cities look when wires are placed underground!

PLATE 88—Birksgate, Glen Osmond, was a fascinating mansion, profusely ornamented by cast iron. The large central part of the verandah was stepped forwards from the verandah balcony. A house of forty-two rooms, it had a large glass and cast iron conservatory and had a plant which made gas for heating and lighting. A 500,000 gallon water tank was installed. Its size, and difficulties of staffing and upkeep of such a house, doomed it to demolition when its grounds were subdivided as a housing estate.

The National Trust of Australia (South Australia) provided the following historical details: "The land on which the house is built belonged to Captain E. C. Frome who reached South Australia in 1839 as South Australia's third Surveyor-General.

"In 1845 Captain Frome sold it to Arthur Hardy (1816–1909) who arrived in South Australia in the *Platina* in 1839. Arthur Hardy was a lawyer and was for a time Crown Prosecutor; he was the son of Thomas Hardy of Birksgate, Yorkshire, and he called the place after his old home. Arthur Hardy built in 1851 the nucleus of the present house and opened the quarries further up the Mount Barker Road past the house.

"In 1864 Arthur Hardy sold the house and surrounding land to Sir Thomas Elder, but not the quarries. Elder however bought further land near the house and very greatly increased the size of the house itself, the main additions built for him being in 1871.

"Sir Thomas Elder (1817–1897) reached Adelaide in 1854 to join his brothers in the firm of Elder and Co., later Elder Stirling and Company, and finally with his brother-in-law Robert Barr Smith in the firm of Elder Smith and Company.

"At Sir Thomas Elder's death in 1897 he bequeathed Birksgate to his nephew Tom Elder Barr Smith, who lived there for many years and himself added to the house, although much of the present house was built for Sir Thomas.

"The house has remained in the Barr Smith family, passing on T. E. Barr Smith's death to his son Sir Tom Barr Smith, and on his death to his son Tom Elder Barr Smith, the present owner. The notable decorations in the drawing-room are by William Morris and Company."

PLATE 89—One of the ornate bows on each side of the front verandah of Birksgate, which would act as sun traps and give shelter from inclement winds

PLATE 90—The balustrade was of a unique delicate pattern. The castings at the top of the bow were registered in Victoria (No. 7, on 11.4.1870 by Wm. Hutchison[26]). The brackets were made of wood

PLATE 91—Benacre, Glen Osmond Road, near Birksgate, is one of the most beautiful houses in Adelaide. Only the brackets are castings

PLATE 92—Boettger's, Ophthalmic Opticians, Observatory House, established 1877. The unusual "look-out," and the business conducted there, were probably responsible for the name of the building. (The balustrade is No. 577, p. 108, in H.C., and was also registered in Queensland, No. 68, 9.4.1888, A. Overend)

A

B

PLATES 93A AND 93B—The Adelaide Arcade, Grenfell Street, a very pleasant early shopping centre, with small shops on two floors on each side of a long light well. Its front is much braver than its counterparts in some other cities. The cresting of the dome is certainly cast iron and it is probable that other decorative elements were also cast. The arcade joins Gays Arcade, which fronts Twin Street, and both were built in 1885. James Cumming was the architect of Gays Arcade and probably of the Adelaide Arcade as well[27]

PLATE 94—Dimora, a large house with massive bow, with planes of ironwork provided by gate, fence, verandah, balcony, and cresting. Large gate-posts, square in section, with filigree decoration, are rarer in Adelaide than in Melbourne. (The balustrade is No. 92, F.C.)

GATES AND FENCES

The gates and fences of Adelaide are highly distinctive. The designs often resemble those of balusters, the palisades of iron rods with cast caps commonly seen in Melbourne and Sydney being rare. The ironwork is usually set upon a stone plinth, and while it may be of the same design as balusters for balconies, it may be taller or shorter than balcony balusters of the same pattern. However, in most instances the fence pattern is bolder and simpler, and the casting stronger than balcony balusters. Massive gate posts of iron, square in section with a filigree design on the four faces, so common elsewhere in Australia, are replaced by solid iron posts of varied design or massive stone pillars. The fences supplement and compliment the ironwork of the house (plate 94). Two of the many patterns illustrated in the Harley catalogue are shown in plate 18, and the complete range illustrated in Fulton's catalogue is shown in plates 95–99. Translated into reality, a number of gates and fences are the subject of plates 100–106.

PLATES 95A AND 95B—Gates and Fences from the catalogue of G. E. Fulton, pages 21 and 22 (courtesy of the State Library of South Australia)

G. E. FULTON & Co., Peel Street, Adelaide. **22**

No. 61.

No. 219.

Scale, ¾ inch=1 foot.

PLATE 95B

G. E. FULTON & Co., Peel Street, Adelaide. **23**

No. 32.

No. 25.

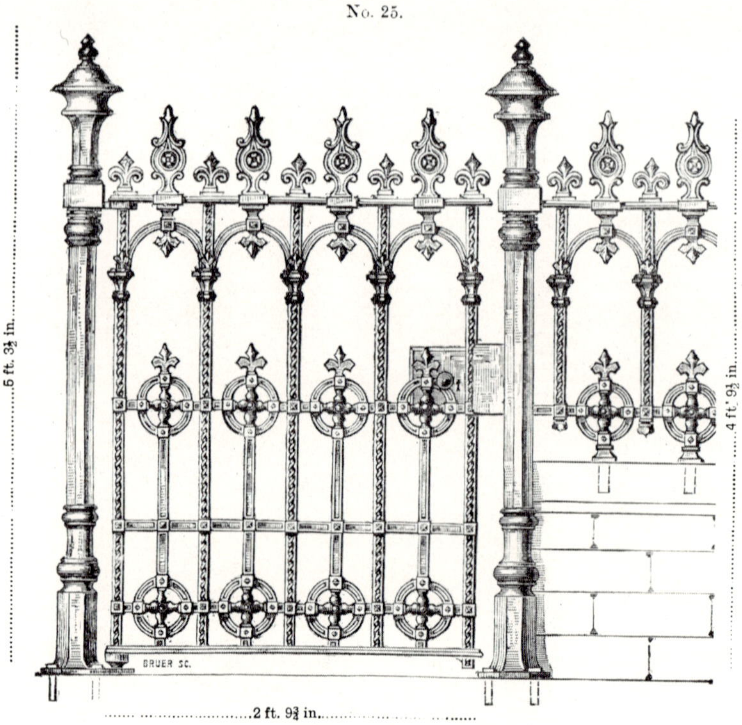

Scale, ¾ inch=1 foot.

Plates 96a and 96b—Gates and Fences, Fulton's catalogue, pages 23 and 24

No. 84.

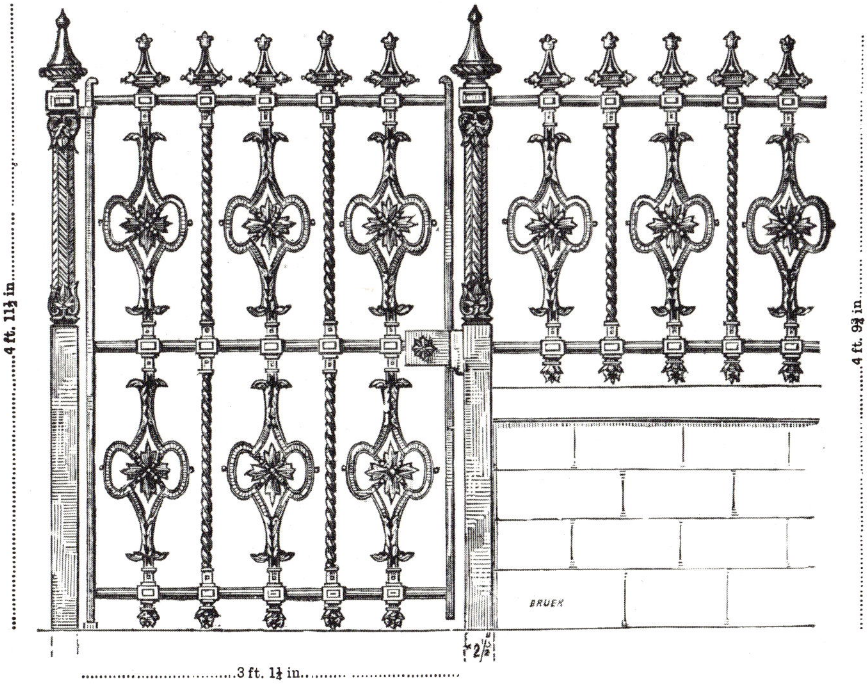

No. 30.

Scale, ¾ inch=1 foot.

PLATE 96B

G. E. FULTON & Co., Peel Street, Adelaide. 25

Scale, ¾ inch=1 foot.

PLATES 97A AND 97B—Gates and Fences, Fulton's catalogue, pages 25 and 26

G. E. FULTON & Co., Peel Street, Adelaide. 26

No. 80.

No. 220.

Scale, ¾ inch=1 foot.

PLATE 97B

G. E. FULTON & Co., Peel Street, Adelaide. 27

No. 203.

4 ft. 3 in.

3 ft. 10 in.

2 ft. 11½ in.

No. 222.

5 ft. 10½ in.

5 ft.

3 ft. 5½ in.

Scale, ¾ inch=1 foot.

PLATES 98A AND 98B—Gates and Fences, Fulton's catalogue, pages 27 and 29

No. 19.

Scale, ¾ inch=1 foot.

PLATE 98B

No. 320.

No. 25.

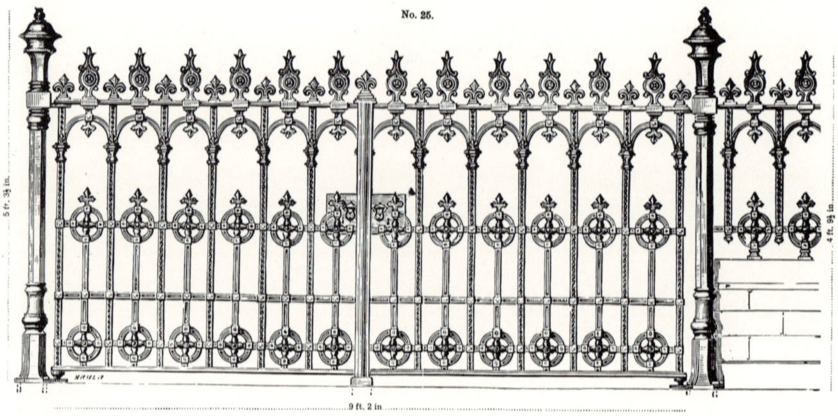

No. 321.

PLATE 99—Carriage Gates, Fulton's catalogue, pages 30, 28, and 32

Plate 100—Iron gate posts, with finials, double-leaved gates, lock, and fence. (The baluster design is shown in Harley's catalogue as a grave fence—No. 800, p. 134)

PLATE 101—Number 17 Barton Terrace. The lily motif dominates the gates, while passion-flower rondels figure above and below. Brackets (of the design shown in plate 31) ingeniously cap the double-leaved gates

PLATE 102—Massive gate posts may sometimes dominate a single-storey house (32 Barton Terrace). The railing pattern was frequently used

A

PLATE 103—The fence and gateway of Cosford, 1 Edwin Terrace, Gilberton
(A) The fence, iron gate standard, and one leaf of the carriage gate
(B) The castings were made for the property, its name "Cosford" being cast on some of the panels

B

Plate 104—A pleasant house, iron fence and gate with massive pillars, and two hitching posts: a North Adelaide landmark at the corner of Hill and Molesworth Streets

Plate 105—Tradition has it that the hitching posts were cast in Austria. However, they closely resemble the type seen in the Deep South of the United States of America, and the physiognomy would correspond. The expression of subservient supplication is skilfully represented

PLATE 106—The cast iron insigne of the City of Adelaide on the Albert Bridge, Frome Road. The delightful vale of the Torrens River is seen through and above the balustrade

2

EVIDENCE OF THE PROVENANCE OF ORNAMENTAL CAST IRON IN ADELAIDE AND OTHER AUSTRALIAN CITIES

IT IS OFTEN SAID that our cast iron ornamentation was brought to Australia as ballast in sailing ships. Some castings were undoubtedly imported from England, but they would need to be carefully packed to withstand breakage during the voyage. Cast iron bears weight magnificently, but thin castings shatter to lateral blows. It is of interest to consider in some detail the provenance of the decorative cast iron used in Adelaide, and in order to do this evidence will also be presented from other States.

The preceding plates and legends show that most of the decorative cast iron of Adelaide can be traced to local manufacture, and indeed much of it came from Fulton's Foundry and the "Sun" Foundry. Other foundries registered designs in South Australia, and their work can occasionally be recognised. Plate 107 shows an amusing example of local design—an iron top hat, in the possession of Mr A. W. Forwood. It was cast by his grandfather, in the foundry of Forwood, Down and Co. (successors to A. C. Harley and Co.), for use as an outdoor ash tray at the time of his daughter's wedding.

An example of importation is provided by the entrance gates to Vailima, Hackney Road (plate 108), on which appears the manufacturer's name, Yates, Haywood and Co., 95 Upper Thames Street, London (see end-papers).

The gateway and fence of the Memorial Hospital, Sir Edwin Smith Avenue, are shown in plate 109. The massive gate standards display the legend, "G. E. Fulton Ltd./Makers/Adelaide." The gates and posts are illustrated in Fulton's catalogue (No. 321, page 32—reproduced in this volume as plate 99c) and in Harley's catalogue (No. 903, page 65), where the fence also appears (No. 876, page 42). It came as a surprise to find the gates, pillars, and fence in the catalogue of the Saracen Foundry of Walter Macfarlane and Co. of Possilpark, Glasgow[28], which illustrates seven massive gate posts, three of which are shown in plate 110. Number 717 (right) corresponds with that shown in plate 109, while No. 745 (left) is to be seen at 261 North Terrace (plate 111), bearing the trade mark Walter Macfarlane and Company, Saracen Foundry, Glasgow.

Plate 112 shows a complete gate with overthrow, which is illustrated in both Adelaide catalogues and in the Glasgow catalogue. This triple listing is not uncommon. This gate and overthrow is widely distributed in Australia, and Macfarlane's name appears on some specimens but not on others, presumably because copies were made in Australia. It is also of interest that the artist's name, J. Bruer, sometimes appears on an illustration in one Adelaide catalogue and not in the other, while the measurements recorded may be slightly different.*

The gate and overthrow are to be seen in Rose Bay, New South Wales. It bears the trade mark of Walter Macfarlane[29] while the same design at 997 Burke Road, Upper Hawthorn has no indication of the maker's name. Some of the elements are seen separately in Adelaide. A side entrance to the South Australian Hotel in Wood Sons Lane had the overthrow and posts, the latter bearing the legend "G. E. Fulton and Co./Makers/Adelaide."

The drinking fountain and cupola, shown in plates 114 and 115, have also been seen at Stansted, Montfichet, England, and in New South Wales, Victoria, and South Australia. Plate 115A shows the complete Macfarlane's fountain, a reproduction of which heads an article in *The Australian Town and Country Journal* of 9 July 1870, which documents its importation into New South Wales[30]:

DRINKING FOUNTAINS FOR THE CITY

> At the suggestion, or order of the Mayor, eight highly ornamented cast iron drinking fountains have been imported by Mr A. Chisholm, from Glasgow, where they were manufactured by Messrs. Walter Macfarlane and Co., of the Saracen Foundry. The fountains stand about eight feet high. The structure, which is pagoda-shaped, consists of four columns, from the capitals of which consoles with griffin terminals unite with arches formed of decorated mouldings, encircling ornamental shields bearing the City Arms, a shield charged with a ship and beehive and working bees, surmounted by a mural crown impaled with a pick-axe surmounted by a star. The sinister supporter is an aboriginal native with a spear; the dexter supporter an English sailor. Motto, "I take, but I surrender." On two of the sides provision is made for receiving an inscription, whilst on the other two sides is the useful monition "Keep the pavement dry." Surmounting this is an open and highly enriched dome, the apex being occupied by a crown. Under the canopy stands the font with basin about two feet in diameter.
>
> It has been suggested that if practicable one of these fountains, an engraving of which is appended, shall be placed in each ward of the municipality, in that part which constitutes the greatest thoroughfare. It is likely that they will be erected at or near the following places: Alfred Park, Flagstaff Hill near the Observatory, junction of William and Macleay streets, the entrance to Moore Park, the Haymarket, near the intersection of Park and Elizabeth streets, near the corner of Market

* Documentation: H.C., No. 894, p. 50; F.C., No. 226, p. 31; Macfarlane's catalogue, No. 845, p. 332. The artist's name appears in both Australian catalogues, but not in Macfarlane's catalogue. The measurements differ slightly in the three catalogues.

> and Sussex streets, and on the Circular Quay near the Custom House. The erection of these handsome fountains will be of general utility, and they will have a very pleasing effect. Their entire cost is £269 3s. 8d.

There is no maker's name on the fountain in Kensington, Adelaide (plate 115B), North Melbourne (plate 114), and Green Park, Sydney, while that in the Commonwealth Gardens, Williamstown, Victoria, bears a diamond-shaped Macfarlane trade mark. The central water basins shown in Macfarlane's catalogues (plates 115A and 116) are characteristic of the imported fountains. Those lacking the Macfarlane trade mark have simpler aqueous arrangements, and it is reasonable to suppose that locally made specimens were supplied with less costly castings. That shown in plate 114 bears the name Danks, a Melbourne firm. A hexagonal modification in the City Park, Launceston (plate 117) shows the name Walter Macfarlane, Glasgow, on the base of each column. The central part alone forms the base of a column previously at the foot of St John's Street, Launceston, and now in the grounds of Franklin House, a National Trust (Tasmania) property. The Macfarlane name is to be seen above its four basins (see end-papers).

Plate 107—The gardener's friend, a local casting

PLATE 108—The carriage gates and lanterns of Vailima, Hackney Road, Hackney, which were cast by Yates, Haywood and Co., 95 Upper Thames Street, London

PLATE 109—The carriage gates and fence of the Memorial Hospital, Sir Edwin Smith Avenue, North Adelaide. The massive gate standard bears on its base, in a raised oval, G. E. Fulton Ltd/Makers/Adelaide

PLATE 110—The gate standard used for the Memorial Hospital is shown in Macfarlane's catalogue (No. 717, p. 339). Number 745 shows the gate posts used at 261 North Terrace, Adelaide (plate 111)

PLATE 111—The iron gate standards and fence which front the Ionic columns of 261 North Terrace (compare plate 110). They bear the Macfarlane trade mark. (The fence also appears in the same catalogue, No. 809, p. 260. A. C. Harley and Co. show a modification of the same design, No. 225, p. 45)

PLATE 112—Cast iron wicket gate and standards, illustrated in the Macfarlane, Fulton, and Harley catalogues

PLATE 113—The gate at O'Sullivan Road, Rose Bay, New South Wales, cast by Macfarlane and bearing his name. Similar columns supporting the overthrow in Wood Sons Lane, Adelaide, bear the legend "G. E. Fulton/Makers/Adelaide"

PLATE 114—An example, in North Melbourne, of a widely used drinking fountain, some specimens of which were cast at Macfarlane's Saracen Foundry, while others have no trade mark. This fountain was locally cast as it does not have Macfarlane's trade mark. The water fountain is different from that shown in plate 115, and bears the name "Danks"

A

PLATE 115A—The fountain illustrated in Examples Book of Macfarlane's Castings, Glasgow and London, not dated, page 40

PLATE 115B—The cupola of the fountain at the corner of Portrush Road and High Street, Kensington, South Australia

B

PLATE 116—Another fountain depicted in William Macfarlane's catalogue. The component parts of the octagonal fountain are the same as those of the square fountain

PLATE 117—The fountain in the Launceston Public Gardens, Tasmania, commemorates Queen Victoria's Diamond Jubilee, 20 June 1886, having been presented by the school children of that city. It is exactly the same as that shown in plate 116, and has Macfarlane's trade mark at the base of each column

PATTERNS ABROAD AND IN AUSTRALIA

Plates 118–126 show some local patterns which are also seen abroad, and in catalogues published in Scotland, France, and Australia. Plates 127 and 128 show the extraordinary façades of the Saracen and Sun Foundries of Glasgow, which will be mentioned frequently[31, 32].

PLATE 118—A fence in Hobart, Tasmania, resembles No. 417 in plate 107 of the catalogue of the Sun Foundry of Geo. Smith and Co., Glasgow. Close inspection reveals certain differences. Another modification may be seen in Cheltenham, England

A

B

C

PLATE 119—A pattern commonly found in London and Melbourne, and less commonly in Sydney and Adelaide

(A) 119 Drummond Street, Carlton, Victoria

(B) Detail of balustrade of 119 Drummond Street, Carlton, Victoria

(C) The design as shown in the catalogue of McDowall, Steven and Co. of Glasgow[33]

PLATE 120—A page from the catalogue of Barbezat et Cie, Paris, published in 1850

PLATE 121—The two designs, united in number 28, plate 120, each form two different balustrades in Sydney

PLATE 122—Two united patterns are shown in the catalogue of the Saracen Foundry of Walter Macfarlane, Glasgow (No. 990, p. 278). They appear together in the gate supports shown in plates 112 and 113, and separately in both Fulton's and Harley's catalogues. They are each used separately in Hindley and Rundle Streets, Adelaide (see plates 48 and 54)

PLATE 123—Numbers 117 and 299 in Macfarlane's catalogue are used in Adelaide. (No. 117 was registered as S.A. No. 21, see plate 27A. It was illustrated as No. 565 in Harley's catalogue, plate 6 in this book, and shown in use in plate 30. Number 299 is the same as No. 132 in Fulton's catalogue and No. 567 in Harley's catalogue, also shown in plate 6 in this book. It is frequently used in Adelaide)

A

PLATE 124—The pattern depicted as No. 299 in plate 123 has been seen in places as far apart as Thailand, Tasmania, London, Sydney, and Adelaide
(A) A delightful small bandstand in Bangkok
(B) An early building preserved as the Department of Agriculture at Deloraine, Tasmania. Such columns were cast in Sydney by the firm of Taylor and Wearing (see *Sydney Lace,* plate 104), and presumably the matching balusters were cast by the same firm. The balusters linked the Ionic columns on the back verandah of demolished St Malo, Hunters Hill, New South Wales

B

PLATE 124c—Number 37 Montague Square, London. Plates 77 and 81 attest to its use in Adelaide. It is similar to a Parisian pattern in Barbezat's catalogue (B.M. in plate 120 in this book). Not infrequently a basic design was modified by other foundries

PLATE 125—Number 178 Stanley Street, North Adelaide, a fine example of Victorian eclectic exuberance. (The spandrels are shown in Fulton's catalogue, No. 159, p. 43, and in Harley's catalogue, No. 938, p. 85, while the cresting above the bow window appears in Harley's catalogue, No. 647, p. 118)

A

B

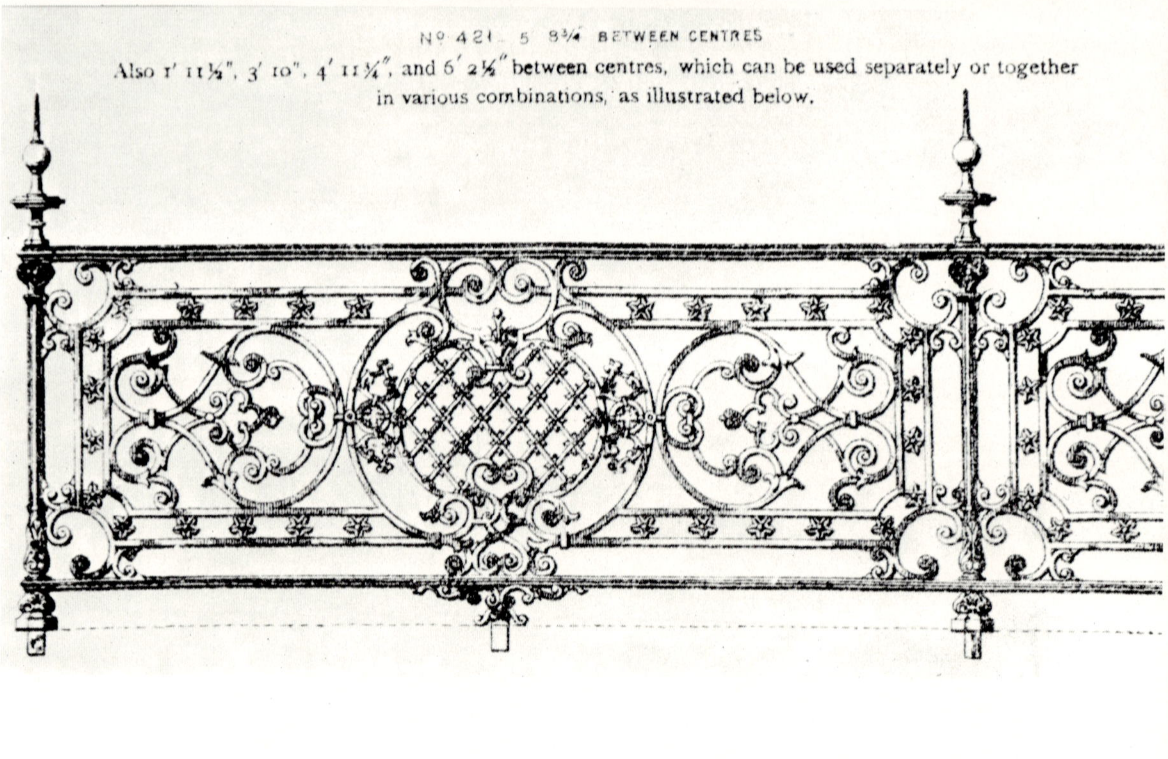

PLATE 126—(A) The wide casting on the verandah of 178 Stanley Street, North Adelaide. The casting is shown in both Fulton's and Harley's catalogues (see plate 3 in this book—No. 960)

(B) A similar casting is shown in Macfarlane's catalogue. The only difference is in the small median floret above the lattice-work. (The castings on 178 Stanley Street are of the Fulton-Harley type and do not carry Macfarlane's name, although the rectangular elevation of Macfarlane's trade mark, shown in plate 129, may be discerned at the base of the column shown in plate 125. It is of interest that the castings on the fence bear the legend "Registered by D. Garlick July 1st, 1884"—see end-papers)

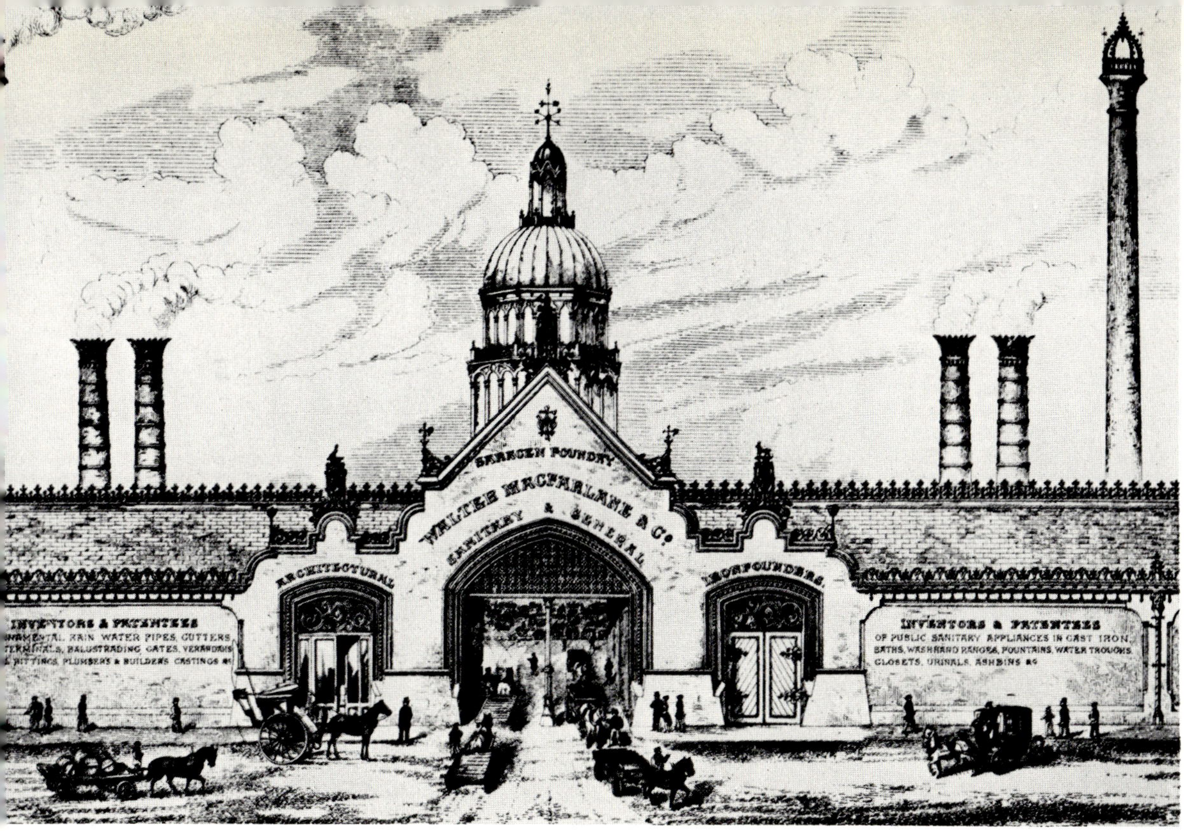

PLATE 127—The Saracen Foundry of Walter Macfarlane and Co., Glasgow[31]

PLATE 128—The Sun Foundry of Geo. Smith and Co., Glasgow[32]

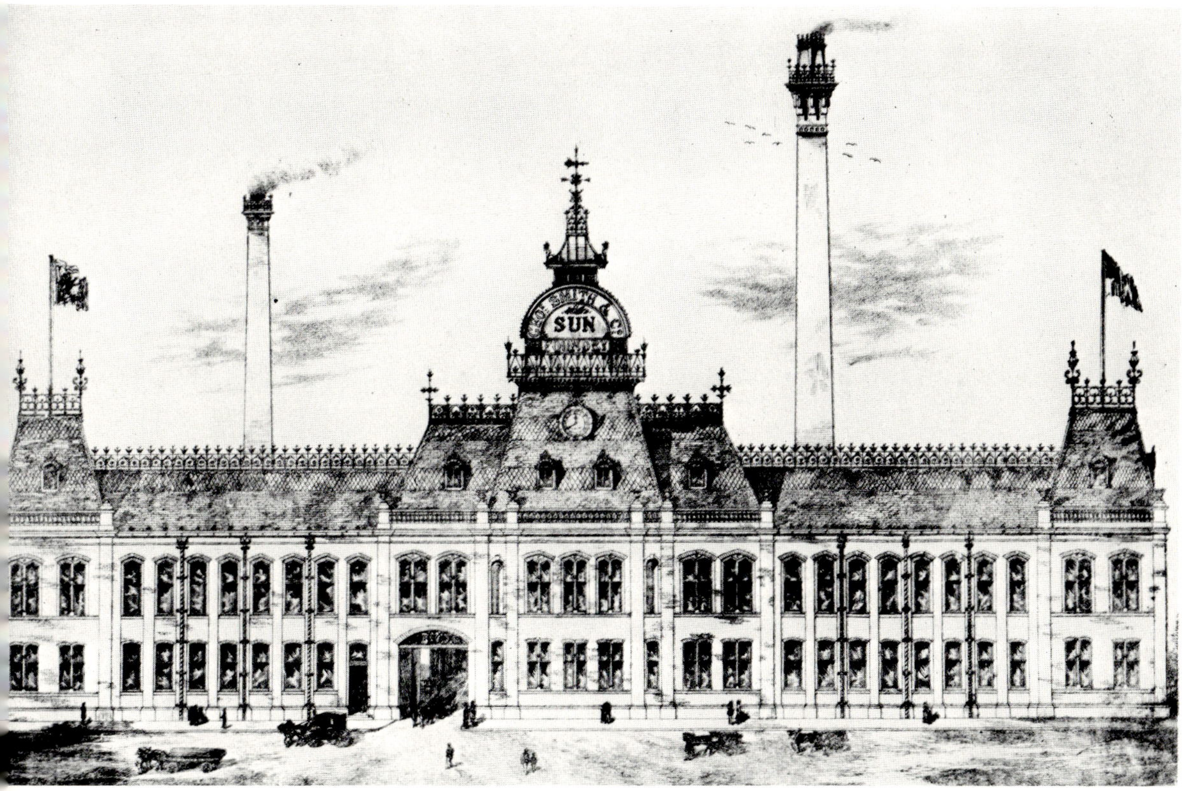

TRADE MARKS

Number 178 Stanley Street, North Adelaide, has wide balusters which were depicted in Fulton's, Harley's, and Macfarlane's catalogues. The base of the verandah columns bears a diamond shaped elevation which suggested the trade mark of Walter Macfarlane, confirmed when the owner, Mr R. F. Hamilton, kindly removed the concealing paint. This trade mark (plate 129) heads the preface to the catalogue of Macfarlane's Saracen Foundry[34], which contains the statement: "We find it necessary to state that every casting has our own Trade Mark legibly impressed upon it." The firm was protecting itself against the copying of its designs.

PLATE 129—The trade mark of Walter Macfarlane and Co., Saracen Foundry[34]

The question therefore arises whether Australian firms copied the designs of Macfarlane, perhaps with permission by right of purchase, or perhaps omitting these formalities. The identical copying suggests that imported specimens were used as patterns for moulding. The key to the solution may lie in the words "trade mark," which, when properly registered, prohibited the application of the mark to any object made by another firm, but did not prevent copying the object itself. Strange as it may seem, many laws and law suits were necessary to define the proper type and wording of a trade mark suitable for registration.

The *Encyclopaedia Britannica* begins its informative article on trade marks as follows[35]:

> TRADE MARKS. A "trade mark" may be defined as a symbol, consisting in general of a picture, a label, or a word or words, applied or attached to the goods of a trader for the purpose of distinguishing them from the similar goods of other traders, and of identifying them as his goods, or as those of his successors, in the business in which they are produced or put forward for sale. A trade mark differs in its legal character both from a patent and from a copyright. In the case of a trade mark the property and the right to protection are in the device or symbol adopted to

designate the goods to be sold, and not in the article which is manufactured and sold. The article is open to the whole world to manufacture and sell, and all that the owner of the trade mark is entitled to prevent is such use of his mark by other traders as will lead purchasers to buy, as his, goods which are not his. On the other hand, patent-right and copyright protect the substance of the article; and any unauthorised manufacture of it in the former case, or reproduction of it in the latter, while the protection lasts, is prohibited. The grounds, however, on which trade marks, patent-rights, and copyright obtain legal recognition, though they are to a certain extent dissimilar, have a common element. Patent-right and copyright rest upon the view that the results of the original labour of the inventor and the author ought, as a matter alike of justice and of public policy, to be secured against piracy; while, as regards the proprietor of a trade mark, the question of originality does not arise so long as the mark is sufficiently distinctive really to identify his goods and, for purposes of registration, to satisfy the Trade Marks Acts.

Signatory states to an international convention in 1883 agreed to an extension of trade mark law to their countries. The Australian colonies had their own trade mark laws—the earliest being Victoria in 1890—which generally followed the English Trade Marks Acts of 1883–88. The Australian Constitution Act of 1900 placed the matter in the hands of the Federal Parliament.

COPYRIGHT

Some Australian castings bear the maker's name. The tall massive columns of the verandah of the South Australian Hotel bore the legend "Stewart/and/Harley/Adelaide," or "Stewart/and/Harley," while the columns of the adjoining shop at 154 North Terrace, have the later name "A. C. Harley/and Co./Makers/Adelaide." Some of these are shown in the end-papers. However, this was not a trade mark in the proper sense of the word, and did not prohibit copying. Registration of designs for copyright purposes was necessary to effect this. The Harley catalogue contains patterns registered by other firms* which had thus sought to protect their patterns.

* The identical designs of the following may be noted:

Design S.A. 21 (Revell, Adams) and No. 565, H.C.
S.A. 32 (D. Garlick) and No. 325, H.C.
S.A. 36 (G. E. Fulton) and No. 895, H.C.
S.A. 42 (G. E. Fulton) and No. 560, H.C.
S.A. 43 (G. E. Fulton) and No. 965, H.C.

Similarities were also noted between designs patented by other firms in other States and in designs in Harley's catalogue, e.g.

Design Qld. 68 and No. 577, Stewart and Harley catalogue (1st ed.)
N.S.W. 93 and No. 961, H.C. (2nd ed.)
Vic. 70 (patented 22.8.1872) and No. 957, H.C.
Vic. 110 and No. 576, H.C.
Vic. 338 (fern pattern) and No. 580, H.C.

Other examples could be given. The Harley catalogues and the Stephens catalogues of Melbourne also show a number of identical patterns, e.g. No. 592, H.C., and No. 20, Stephens catalogue; No. 561, H.C., and No. 29, Stephens catalogue.

A law suit, joined in Sydney on 16 February 1887, throws considerable light on the registration of an article for copyright purposes. The trial is reported in considerable detail in *The Australian Ironmonger*[36].

In the Equity Division of the Supreme Court of New SouthWales, Swinnerton and Frew, iron-founders, sued William L. Dash, an iron-founder and contractor, for infringement of copyright. The latter had been granted to the plaintiffs under the seal of the Office of Copyrights of New South Wales for the space of three years from 23 November 1885. Two designs were involved: New South Wales registered numbers 217 (a balcony railing with ferns worked in iron), and 218, a frieze and bracket in iron, with floral devices and ornaments. The defence was that the registered designs of the plaintiffs were published in the Colony of Victoria previous to such registration, by other persons than the plaintiffs, who were not the originators of the designs, and had no legal right to the copyright.

On 5 August 1886, Alfred Swinnerton had inspected ironwork being carted to a building, and found it to be exactly of his design. There was no registration mark on it, it had been cut away probably in the mould, but there was a rough mark on the bar where the name of the plaintiff's firm should have been. It appeared as if the words had been obliterated in the pattern. His firm had not manufactured this ironwork.

Questioned by the admirably named Dr Sly, he admitted that he was not the designer of this ironwork, but had purchased the designs from a Mr Macpherson for five pounds. He believed Macpherson was not the original designer; he did not know from whom that gentleman purchased them.

Dr Sly submitted that the "proprietor" under the Copyright Act should be the author of a new and original design, or should purchase the design directly from the author.

Macgregor Simpson, iron-founder, said that he had visited Melbourne in January 1885, and was offered for purchase a pattern of the same design by C. Monteith and Sons at Sandwich, Victoria. He had not wanted to buy, and no price had been stated.

Hunter Macpherson, iron merchant, stated that he had got the designs from his father in Melbourne, and had sold them to Swinnerton and Frew simply as a casting, not as an original design of his own.

Dr Sly maintained that the Act required, before copyright could be granted, that the articles had not been published (i.e. manufactured and sold in the usual way) in the state of origin or elsewhere. Registered castings could be sold in another state, but could not be used as patterns.

Mr John Monteith of the firm of Charles Monteith and Son, was brought from Melbourne for the adjourned hearing. He stated that the pattern was carved for his firm from a drawing supplied by James Marriott. He held the registration of the design, and was the chief originator of the design, which had been registered and published (i.e. manufactured and sold) in Melbourne. The plaintiff's counsel conceded that this invalidated registration in New South Wales, but his client had believed he obtained a new design and registered it, and had a right to it. He said the plaintiff purchased the designs, and had acted *bona fide* in the matter.

His Honour said that the fact that the plaintiff purchased the pattern from Mr Macpherson at a cost of five pounds did not give him the right of registration in this Colony. It had been shown there had been a previous registration in Melbourne, and the suit was dismissed with costs.

In summary, the castings were registered, produced, and sold in Victoria, previous to the New South Wales registration by the plaintiffs and the use in New South Wales by the defendant. The design was registered in Victoria, under a Victorian statute, and later in New South Wales under a New South Wales statute. The New South Wales statute, however, provided that there should be no copyright of designs which had been previously published in New South Wales or elsewhere. Publication in Victoria was proved, and therefore the New South Wales registration was invalid and ineffective.

* * *

The appearance in Harley's catalogue of designs registered by other firms in South Australia or in other States might be due to the "Sun" Foundry selling castings made by other firms, or by purchasing a design from the copyright holder, or reproducing it after a lapse of three years from the date of registration. The identity of pattern in the Fulton and Harley catalogues may be due to purchase, or agreement, when the Fulton firm went out of business. At a later period, the "Sun" Foundry was using the floor and "boxes" of the old Fulton Foundry for producing its castings.

COTTINGHAM'S *DIRECTOR* AND IMPORTED DESIGNS

It was fashionable for other artistic crafts to publish pattern books, which served as a basis for manufacture by others, and to educate the taste of the public in general. A number of such pattern books dealing with wrought and cast iron were published, and the most popular of these was prepared by an architect, L. N. Cottingham, in London. The work appeared in two editions. *The Ornamental Metal Worker's Director* in 1823, and the second edition named *The Smith and Founder's Director* in the following year. The preface of the first edition is illuminating[37]:

> The extensive application of metal, in securing, decorating, and furnishing every class of building, from the superb palace of the monarch to the social villa of the retired citizen, renders any apology for introducing a Work of this description unnecessary. The great improvement that has taken place in our Brass and Iron Foundries within these last twenty years, has elevated this branch of English manufacture far above that of any other country, and raised the articles which were formerly considered as merely gross and ponderous, into the scale of ornamental embellishment, in which utility and security are united with the lightness and elegance of classical design. The fidelity and correctness with which the Smiths and Ornamental Metal-workers execute their orders, must make a work, which is calculated to improve their taste, and excite emulation in getting up their patterns, generally useful and beneficial; and the decided favour that has ever been shown by gentlemen of fortune and liberal minds, to the ingenious artisan and aspiring tradesman, who are capable of combining utility and elegance in the various articles for domestic purposes, has been a sufficient stimulus to the Author to use every means in his power to introduce such a collection of Designs and Patterns as may be a guide to them in forming correct and tasteful compositions, the only way to insure a preference for the British manufacture in every class, and prevent the inundation of foreign goods, which have long obstructed the rising fame of our artists in the higher departments of this art.
>
> Many of the subjects introduced in this Work have been executed from the designs of the most eminent artists; and those composed by the Author are from the best specimens of antique ornament, consisting of Entrance Gates to public buildings, parks, gardens, &c. Verandas, Fences: Balcony, Area, and Window Guards; Balustrades and Newels for staircases and galleries; Fan-lights, Lamps and Brackets for entrance doors; Street Lamps; grand Stands for gas-lights; Tripods, Candelabra, Candlesticks, Chandeliers; Vases and Pedestals; Hot-air Stoves for churches, chapels and public offices; elegant Stoves and Fenders for drawing rooms, &c &c. serviceable for ladies and gentlemen to select from, and equally so to the furnishing Ironmonger. To remove in some measure the severe and painful regret that has long been felt by

ingenious workmen, for the want of a collection of good ornaments to select from, at a price within the compass of their limited means, the Author has given several hundred specimens of the choicest productions of the Grecian, Etruscan, Roman, and Gothic schools of art; from accurate drawings and casts in his possession, which he flatters himself will tend to facilitate the operations of the professional artist, and afford ample matter for the mechanic to study from, to whom it will be excellent practice to draw the smaller ornaments three or four times the size given in this work.

L.N.C.

66, Great Queen Street, Lincoln's Inn Fields,
October 10th, 1823.

These designs were published for copying. Those which appeared in Australia could be cast locally or imported from England. In one instance (No. 1, plate 130) importation was documented. A fine example of Regency architecture, Hythe, was built for William Pritchard Weston, who migrated to Tasmania from London in 1824. The architect of Hythe was S. Jackson, who later went to Port Phillip with John Batman, and was responsible for many early Melbourne buildings. He wrote a letter to Mr Weston dated 19 August 1831 (now preserved among the Weston papers in the Launceston Museum) suggesting that two complete sets of ironwork of the lightest possible weight, should be ordered from England.[38] The resulting ironwork, the pattern of which is to be found in L. N. Cottingham's *Smith and Founder's Director*[39], contributes to the beauty of the house (plate 131). The same pattern is to be seen elsewhere in Tasmania, in other Australian cities, and in London (plate 132). Another of Cottingham's designs[40] (No. 8, plate 133) was frequently used in London, and also appears in Australia (plate 134). A casting of Royal Arms, seen in England, Australia, and India, is shown in plate 135. Another baluster pattern, probably used early in Sydney's development[41], is also to be seen in London (e.g. in Brompton Square). All the former castings may have been imported into Australia or reproduced in bulk in Australia from earlier imports, while early and uncommon types were probably imported.

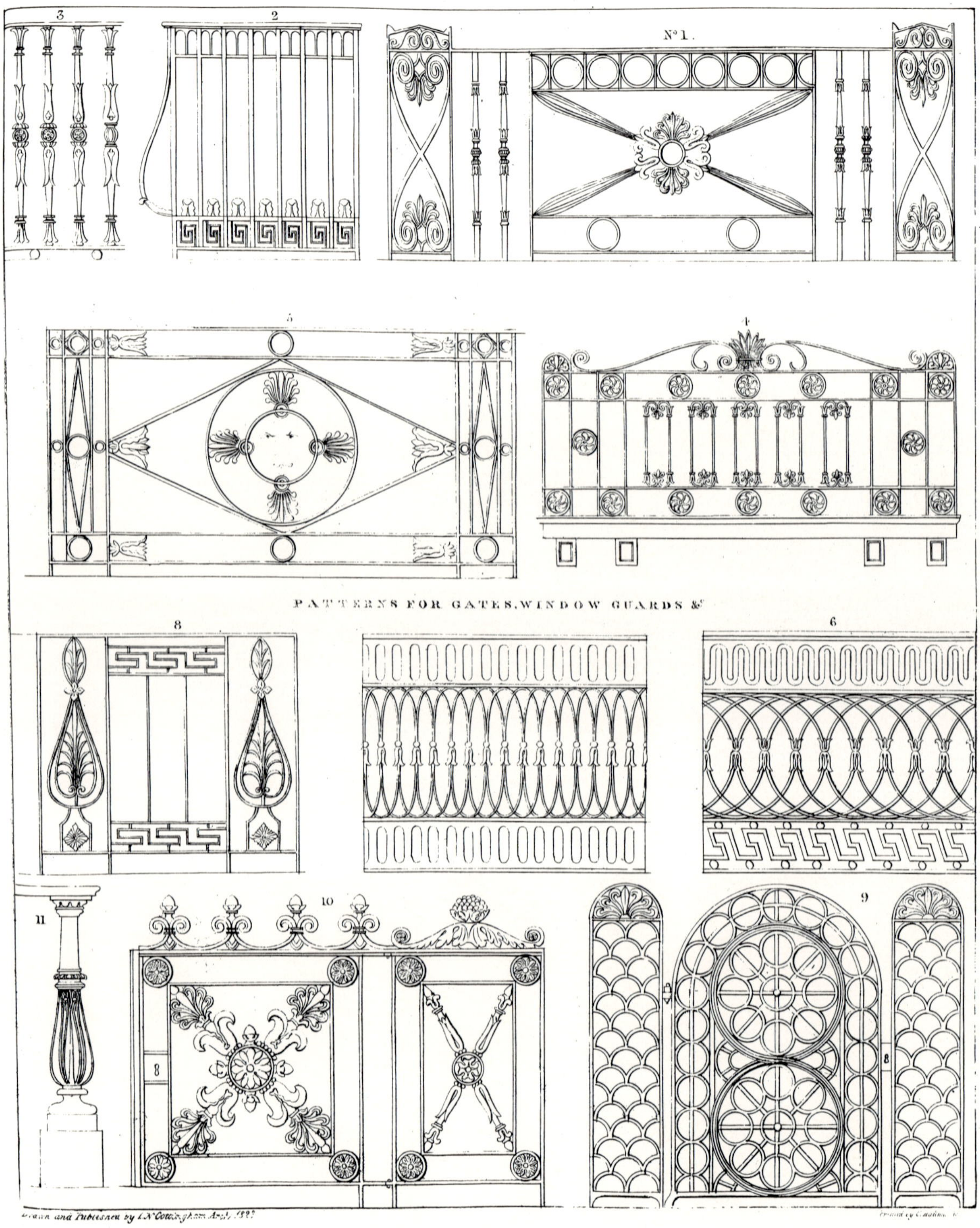

PLATE 130—Plate I from *The Smith and Founder's Director* (1824) by L. N. Cottingham, Architect[39]

PLATE 131—Hythe, Tasmania, a delightful house, the Regency characteristics of which extend to the ironwork. The balusters are shown as No. 1 in Cottingham's *Director* (plate 130). This is one of the few instances in which the purchase, in England, of the ironwork is documented in correspondence

PLATE 132—The same pattern is not infrequently seen in London (Amwell Street, Finsbury)

PLATE 133—Plate II from L. N. Cottingham's *The Smith and Founder's Director* (1824). Number 8 is another pattern common to London and Australia[40]

PLATE 134—The pattern, No. 8, plate 133, is to be seen in London (Myddelton Square), and in Sydney (171 Macquarie Street)

PLATE 135—Royal Arms, from the catalogue of A. C. Harley and Co., Adelaide. The same design was cast by Walter Macfarlane. A specimen (138 x 82 cm) stored in the Prince of Wales Museum of Western India, has Walter Macfarlane and Co., Glasgow, cast on the horizontal quartering. It differs from the Royal Arms cast by P. N. Russell of Sydney (plate 139)

LOCAL PRODUCTION

After dealing with register grates, the entry of Messrs Fletcher Brothers Foundry, Camdenville, New South Wales, in the *Catalogue of the Sydney International Exhibition* of 1879 states:

> Even in those articles which are directly copied from the English patterns, the advantage on the side of Sydney as compared with English wares, is very marked. Every description of ornamental balconies, friezes and pilasters are turned out of the factory, these being chiefly of Messrs Fletcher Bros. own designing. The patterns are really artistic and in no line is the taste of the workmen shown to greater effect than in the many varieties of fluted columns which are constantly under order. The fact that so large a quantity of the last-mentioned work is turned out at the Camdenville foundry is due to the corporation having enacted that all persons making improvements in house property within the city shall use one or other of two given designs, instead of, as formerly, having full liberty to erect whatever kind of ironwork they pleased.

Messrs Taylor and Wearing's Foundry of Redfern, New South Wales, stated:

> The firm have a good assortment of plain and fluted column patterns, spandrills, etc, of which work they turn out a considerable quantity.

The Central Foundry, Clarence Street, displayed:

> Wrought-iron railings for flagstaff hill, and portions of the railings around Hyde, Cook and Moore Parks, all of which works were executed in a manner highly creditable to the firm. Specimens of the ornamental gates made at the Central Foundry may be seen at the warehouses . . . Pilasters and often ornamental architectural castings are extensively dealt in by this firm . . .

The most interesting and informative casting in South Australia is a massive fence post fronting the Gawler Institute at Gawler (plate 136). It proclaims the smelting of ore raised in the Barossa Valley, and the manufacture of the post at the Phoenix Foundry, Gawler, in 1871, from the first iron smelted in the Colony. It was presented by James Martin, iron-founder, in 1879. The name J. Martin appears on the cast iron latch of a gate in Murray Street, Gawler (see end-papers). The latch also names place, Gawler, and date of manufacture, 1874. It is sad that this precise documentation is so rare.

The Gawler Handbook by George E. Loyau, published in 1880, provided the following information of the iron master, and incidentally gave first hand information of the production of castings in Australia[42]:

> Mr James Martin, originator of the Firm, is one of the most conscientious and honourable of men; esteemed by all who have the pleasure of

> his acquaintance, he goes on year by year adding to his friends and decreasing his enemies. Few have persevered as he has done, and thereby won the reward due to honest toil . . . He left Cornwall, of which he is a native, in his 25th year and arrived in Adelaide July 1st, 1847 . . . He worked in the metropolis for eleven months, when he went to Gawler, and hired a blacksmith's shop from Mr Henry Calton. There was only one blacksmith's shop with one forge and one man working . . . He had uphill work at starting, having but little money and little credit . . . First difficulties being surmounted, he grew prosperous, and in 1851 was able to employ as many as thirty men, when a rush took place to the diggings. As everybody went, he joined some companions but was unsuccessful in the search for the precious metal. Returning to Gawler he resumed work.

The *Handbook* also provides information of the Eagle Foundry, its proprietor, Mr D. Thomson, and of the smelting of iron ore in South Australia[43]:

> The first object in the establishment to which attention is directed is the raw iron, which is in pigs* weighing 1 cwt. each. These are imported from England, and are considered of excellent quality . . . There is plenty of iron ore in South Australia, and experiments made from it prove its utility and adaptability for manufactures. The cost of reducing it from its crude state is, however, too expensive, and from the absence of coal is a sufficient deterrent to prevent its being used for some time to come.

A photograph of a settee, "Four Seasons," cast by D. Thomson, is included in the *Handbook* (plate 137). This design has been used on settees and sets of four chairs in many parts of the world. Exportation from infant Australia was hardly likely. The costumes suggest a European origin. An advertisement for the foundry is also to be found in *The Gawler Handbook* (plate 138)[44].

It is sad to record that Mr J. Martin's attempts at the primary smelting of iron ore were a commercial failure. Among Miscellanea in the *History of Gawler* the following appears (page 351)[45]:

> From the "Bunyip" of November 9, 1872:—This week we have the highly gratifying intelligence to communicate that Messrs Martin & Co., of the Phoenix Foundry, have experimented successfully upon the smelting of our native ores. We yesterday visited the firm's new foundry and had the satisfaction of seeing turned out of the moulds, of excellent quality, cog wheels, ploughshares, a medallion portrait, looking glass frame, handsome picture frame, and various other articles, including a tea cup with saucer and teaspoon. The fact is thus established that we shall be able to rear up and multiply metallic pigs to any extent.

* The molten iron flowed from the blast furnace and ran into a central channel with multiple parallel short branches on each side, called sow and pigs. The stout bars of pig iron were remelted in foundries to reproduce in moulds the patterns of carved wood or of already made castings. The stout short bars of pig iron would withstand shipment as ballast or cargo, and were imported into Australia before local smelting of iron ore became feasible.

> Messrs M. & Co. are now constructing the necessary appliances for a movement on an enlarged scale; Barossa supplying both the iron ore and the charcoal, and we need hardly say that the realisation of this great desideratum must prove of vast importance to those interests of the colony, which we would not expect to enumerate fully at present if we made the attempt.

Another paragraph, obviously referring to a later date, stated (page 354)[46]:

> The late Mr James Martin hoped that iron ore would be successfully smelted here. There are large deposits of the material, but unfortunately, coal has to be brought from a great distance, and this renders the process an unprofitable one. In the late seventies Mr Martin tried what could be done, but while the smelting was a success the cost was too great.

It is of interest that the first attempts at smelting iron ore in two other Colonies were commercial failures. At Mittagong in New South Wales, in 1848, the first smelting works in Australia was established, and ultimately shut down[47]. A blast furnace set up at Lal-Lal in Victoria in 1875 failed, for similar reasons[48].

The use of iron ore from Mittagong is documented on a plaque at the entrance of the new Peter Nicol Russell Building of Engineering Faculty, Sydney University. The Royal Arms displayed above the plaque shows the skill of the craftsmen responsible for its manufacture (plate 139).

The Russell Foundry is mentioned by Joseph Fowles in his fascinating *Sydney in 1848*[49]:

> At the back of Messrs Harper and Moore's [an ironmonger next to the Royal Hotel in George Street] is the Sydney Foundry, an extensive establishment belonging to Messrs Russell.

The Peter Nicol Russell Building links this early foundry and it proprietor with the present era of engineering[50]. A monument in the forecourt commemorates the benefactor whose gifts in 1896 and 1904, totalling £100,000, made possible the establishment of the School of Engineering. Executed by Bertram Mackennal and entitled *Industry,* it carries a bust of the benefactor, a descendant of the original iron-founder (plate 140). The inscription on the pink granite base reads:

> Presented to the University of Sydney by Dame Charlotte Russell in memory of her late husband, Sir Peter Nicol Russell, a munificent benefactor, who endowed the Peter Nicol Russell School of Engineering, June 23, 1907.

One of the foundry's original cast iron columns has been set up in the forecourt, near the monument (plate 140). The name of the foundry is cast on the base (plate 141), a name which may be seen on many castings in Sydney, including the filigree (open-work) columns so characteristic of Sydney.

PLATE 136—A superbly informative casting

PLATE 137 — The "Four Seasons" settee, cast by the Eagle Foundry of D. Thomson, copied from *The Gawler Handbook*

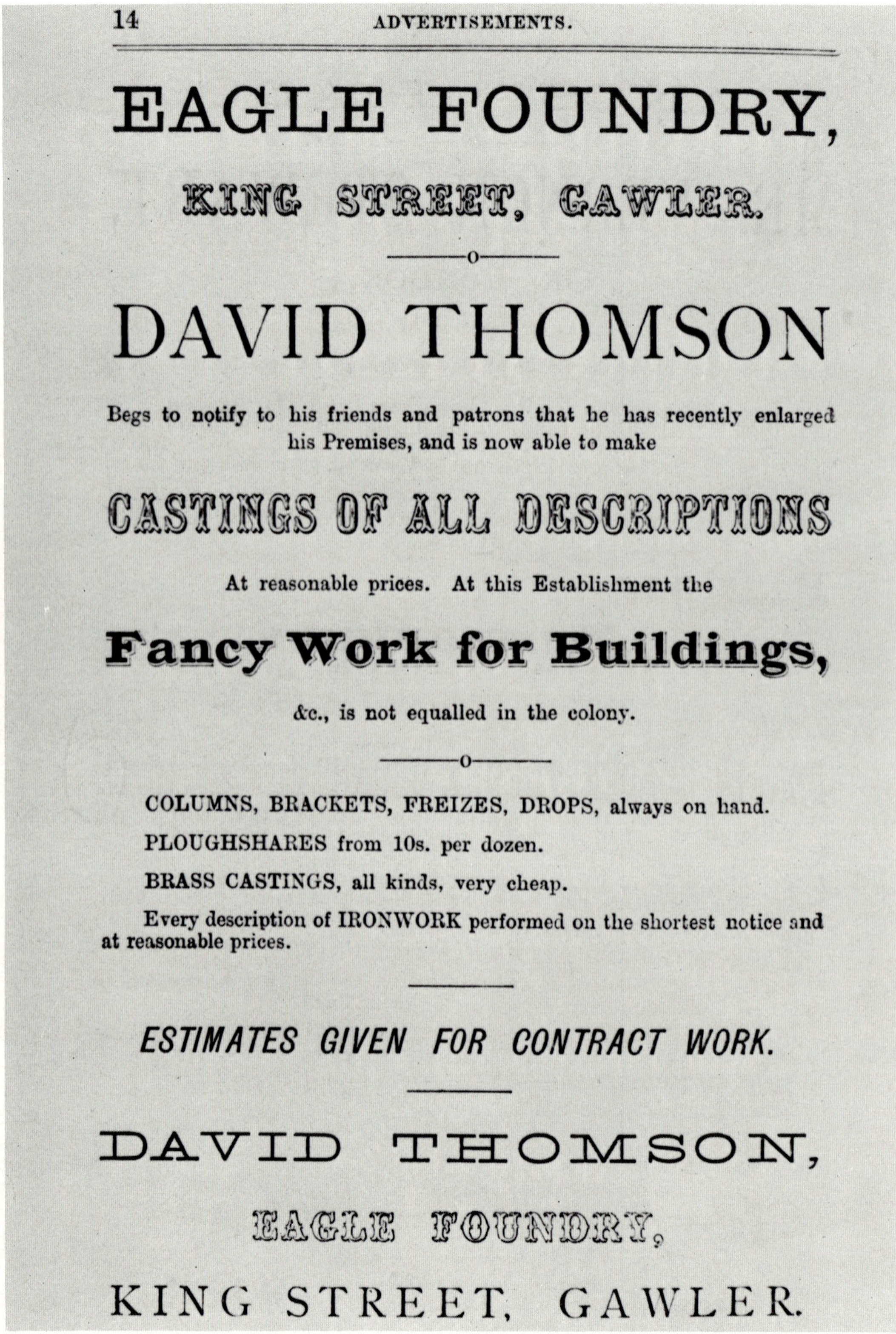

14 ADVERTISEMENTS.

EAGLE FOUNDRY,

KING STREET, GAWLER.

———o———

DAVID THOMSON

Begs to notify to his friends and patrons that he has recently enlarged his Premises, and is now able to make

CASTINGS OF ALL DESCRIPTIONS

At reasonable prices. At this Establishment the

Fancy Work for Buildings,

&c., is not equalled in the colony.

———o———

COLUMNS, BRACKETS, FREIZES, DROPS, always on hand.

PLOUGHSHARES from 10s. per dozen.

BRASS CASTINGS, all kinds, very cheap.

Every description of IRONWORK performed on the shortest notice and at reasonable prices.

ESTIMATES GIVEN FOR CONTRACT WORK.

DAVID THOMSON,

EAGLE FOUNDRY,

KING STREET, GAWLER.

PLATE 138—Advertisement for Thomson's Eagle Foundry, Gawler[44]

PLATE 139—Royal Arms, at the entrance to the Peter Nicol Russell Building of the Faculty of Engineering, University of Sydney. A bronze plate documents its origin: "These Royal Arms were cast by Messrs P. N. Russell's works in Sydney with iron manufactured from ore obtained at the Fitzroy Iron Mines and exhibited at the International Exhibition held in Hyde Park, London, in the year 1851." The casting shows the high degree of craftsmanship attained so early in the history of casting in the colony

PLATE 140—The forecourt of the Peter Nicol Russell Building displays an Ionic column cast by P. N. Russell and Co., and a monument to Sir Peter Nicol Russell, the benefactor of the School of Engineering

PLATE 141—The base of the Ionic column cast by the Russell Foundry

THE EARLIEST COLUMNS CAST IN SYDNEY

Joseph Fowles drew and described the Congregational Church, Pitt Street, for his *Sydney in 1848*[51] (plate 142). In his fascinating description of the church he indicated that the columns (shown in plate 143) were the earliest columns cast in Sydney.

> The Congregational Church, or, as it is frequently called the Independent Chapel, Pitt Street, is the handsomest building of the kind in Sydney, or indeed in any of these Colonies. It was designed by Mr John Bibb, Architect, and erected under his superintendence. The foundation stone was laid by the Rev Dr Ross on the 26th of January, 1841; but in consequence of the embarrassed state of the Colony it was soon after thought prudent to suspend the progress of the work, and it was not until October, 1844, that it was determined to recommence it. The Church was opened for public worship on the 1st January, 1846, by religious services in which various Ministers of other denominations of Christians took a part. The whole cost of the erection has amounted to upwards of £6,000, which has been raised by various contributions, it being contrary to the principles of this body of Christians to accept of assistance from Government, either for the building of their churches or the support of their ministers.
>
> The front of the building is of the Ionic order of Architecture, with two columns twenty-six feet in height, according to the Ionic Temple of the Ilisus at Athens: and four Antae pilasters with ornamental

PLATE 142—The Congregational Church, Pitt Street, Sydney, one of the most interesting buildings in Sydney. The illustration is taken from Fowles's *Sydney in 1848*[51]

capitals; with an entablature and pediment extending over the columns and pilasters. The design is in imitation of the smallest Grecian Temple, denominated by the Greeks a "Temple in Antis" at the principal entrance to the Cella (or interior) of the Temple. The whole front is of polished free-stone, and is divided from the street by an ornamental iron railing on a dwarf stone wall. The pulpit is octagonal, standing on a bold pedestal, and has many carved mouldings of Grecian leaves and ornaments; it is ascended by a winding stair, and is made of the most handsome cedar, which is French polished. The edifice is ninety-five feet long, forty-five feet wide, and thirty feet high, all in clear of the interior. The ground floor has two entrance vestibules, which contain the stairs leading to the galleries, and between them is a room used as a vestry or school room, and which will contain about one hundred and thirty children. Above the vestibules and the vestry is a large room, divided by a baluster railing from the galleries, used as a school room and sittings for the Sunday school children during divine worship; it will accommodate two hundred and forty children. The interior of the Church is elliptical on the ground plan at the end opposite to the pulpit, and three pews in width against the wall run parallel to the sides and circular end with an elevation of four inches at each pew, an arrangement favourable for conveying the sound to all parts of the spacious building; galleries at two sides and at the end opposite to the pulpit follow the same curve as the walls and pews below, and are supported by fourteen cast iron columns of the Ionic order and fluted. These were cast by Mr Dawson, of Sydney, in a very neat and superior manner, and are the first series of ornamental columns cast in the Colony—they are finished in imitation bronze. The Church will accommodate about one thousand persons with sittings, besides the children of the Sunday school. The whole of the pews are of cedar, and the breast-work framing in front of the galleries is of the most choice kind, equal in appearance to mahogany. The place is well lighted with gas, except the pulpit, on each side of which are handsome brass sconces for wax candles.[52]

MAKER'S NAMES ON CASTINGS

It is fortunate that many Adelaide castings bear the maker's name. In contrast, it is rare to find the name of foundries on castings in Melbourne. Some Melbourne columns bear the legend "Maclean's Patent," but these were made of galvanised iron, filled with a tough composition, which were said to be as strong as cast iron, but cheaper. The names of many Sydney foundries were cast on their products[53]. Some further monumental gate posts, bearing the names of the makers, are shown in plates 144–146.

PLATE 143—The interior of the Congregational Church, Pitt Street, which bears comparison with some of London's churches of the Georgian era. The columns supporting the balcony are of Doric type, while those above have Corinthian capitals. The pulpit and its stairs with cast iron balustrades have been altered, for they do not correspond to Fowles's original description[52]. Nor are the columns of Ionic type, this probably being an error on Fowles's part

PLATE 144—Gate posts at Centennial Park bear the legend "Pope Maher and Co., Sydney, Darlington Iron Works"

INDIGENOUS PATTERNS

Indigenous patterns provide further evidence of Australian manufacture, and demonstrate the ability of the designers (plates 147–153). Plates 149–151 are of especial interest. A number of fern designs can be recognised as Australian botanical types. The waratah (plate 152) is exclusively Australian and it is strange that it was not represented more often.

"Corporation brackets," embodying a design representative of a city or town were frequently used on friezes and brackets of shop verandahs. The arms of the City of Melbourne[57] were cast into the brackets of the lofty columns which supported the roof of the Eastern Market, which, unfortunately, has been demolished (plate 153). In fifty years' time it would have been valued as a magnificent period piece, of great interest to visitors. It is high time the citizens of Melbourne took stock of what is happening. The city is losing its grace, once so evident to visitors, and but for its parks, cast iron ornamentation, and a diminishing number of grand buildings, some saved from destruction only by an enormous expenditure of energy by the National Trust, what is there to distinguish Melbourne from other cities? Unfortunately, affluence is counted in terms of cash, not of desirable environment.

PLATE 145—Gateway, 360 Moore Park Road, Paddington, New South Wales

PLATE 146A—Details of gate post of 360 Moore Park Road, Paddington. The solid incised faces are unusual. The posts were cast by D. and R. Bradford, 323 Castlereagh Street, Sydney. In the background, No. 358 is seen. On Plate 146B the abutting smaller fence post of No. 358 bears the legend "Reynolds, Maker, Sydney"

PLATE 146B

PLATE 147—Some indigenous patterns. A frieze and bracket depicting kookaburras—a rare design

PLATE 148—A frieze showing birds in foliage, with a sulphur-crested cockatoo as the central motif of the bracket (registered on 3.12.1884 by D. and R. Bradford, N.S.W. No. 180—the corresponding ornamental balcony railing was registered on the same day, N.S.W. No. 181)

Plate 149—An Aboriginal with weapons features in a design which prominently displays the foundry of origin—Harvey Sargent and Co. (Registered Qld. No. 44). In the interests of propriety the Aboriginal has been provided with a grass skirt more appropriate to South-Sea Islanders

Plate 150—The corresponding bracket shows an emu and kangaroo, as well as the maker's name (registered Qld. No. 46, 9.5.1887)

Plate 151—A balustrade in Brisbane shows alternating narrow and broad balusters. Close inspection shows many additional elements of design. The rose of England, thistle of Scotland, shamrock of Ireland, and the leek of Wales betoken our origins. A palm tree and cactus leaves support the emu and kangaroo and some of the other curvaceous foliage may represent Australian flora. Kangaroo paws support the manufacturer's name

Left

PLATE 152—The waratah *(Telopea speciosissima)*, a uniquely Australian flower, forms the central motif of a cast iron gate illustrated in *The Australian Flora in Applied Art,* by R. T. Baker[54]. It was designed by Parnell Johnston at the Sydney Technical College

Below

PLATE 153—The arms of the City of Melbourne formed the central motif of the huge brackets of the Eastern Market, Melbourne. Taken during demolition of the building

IMPORTATION OF DECORATIVE STRUCTURES

Ornamental castings were also imported from other countries. A massive fountain in St John's Square, Launceston, Tasmania, was imported from Paris (plates 154 and 155). The documentation is cast on the fountain. It was erected by the Launceston Water Works in 1857, having been cast by Barbezat et Cie, Foundries Du Val D'osne, Paris; M. Moreau, Statuaire; Lienard, Sculpteur.

The bandstand in Elder Park, Adelaide, was manufactured by Walter Macfarlane of Glasgow (plates 156–158). Such prefabricated structures, manufactured abroad, are fascinating examples of imported cast iron serving architectural and ornamental purposes. British bandstands have spread around the world, and good examples may be seen in Thailand (plate 124A) and Bermuda.

PLATE 154—Detail of base of fountain, St John's Square, Launceston

Plate 155—Fountain in St John's Square, Launceston, cast by Barbezat et Cie, Paris

PLATE 156—Bandstand in Elder Park, Adelaide

PLATE 157—Detail of the bandstand's balustrade. The Walter Macfarlane name plate is at the base of the columns

PLATE 158—The bandstand is shown in Macfarlane's catalogue[58]

3

PREFABRICATION AND PORTABLE DWELLINGS

SAINT LUKE'S CHURCH OF ENGLAND, Whitmore Square, is a combination of a mutilated ironwork church and masonry (plates 159 and 160). The unusual history is given in the centenary booklet of the Church, kindly supplied by the present Rector, the Reverend G. C. Bennett[59]:

> Divine Service appears to have been first conducted in a Schoolroom in Hobson's Place, between Sturt and Wright Streets. In December, 1853, a meeting was held in Holy Trinity Schoolroom, presided over by Mr Justice Cooper, when it was reported that a sum of £120 had been subscribed for the purpose of erecting a Church in the south-west portion of the City. At that meeting the following resolutions were carried:
>
> 1. That a Church was required to hold 450 at least.
> 2. That Whitmore Square be the site selected.
> 3. That it was desirable to send to England for an iron Church.
> 4. That subscription be collected for same.
> 5. That a Building Committee be appointed.
>
> The Bishop of Adelaide, the Right Rev. Dr Short, who was then in England, was requested to purchase the Church.
>
> In April 1854, a meeting of the subscribers was held. At this meeting a letter was read from the Bishop, stating that as an iron Church would cost £2,000 (exclusive of expenses), he thought it inexpedient to purchase such a Church and advised against doing so. Special meetings were then held and it was decided to proceed with a stone one, and Mr E. Wright was appointed Architect.
>
> Unfortunately, the Bishop, after the despatch of his letter, gave orders for the iron Church to be built and forwarded, and the foundations of the new Church had been put in when the Church arrived. This iron Church, however, on arrival, was condemned by Lloyd's

Agent. To further add to the complication, the Church had not been properly insured; the first monies subscribed were, therefore, practically thrown away. However, the present structure was proceeded with, fresh subscriptions came in, and part of the iron Church was used in the building as it now stands, the remainder being sold.

This might be called the Mystery of the Missing Cast Iron Façade.

A number of prefabricated churches were imported from the United Kingdom, some being manufactured by Hemming's Patent Improved Portable Buildings Manufactory at Clift House, Bedminster, Bristol. The *Illustrated London News* of 30 April 1853, describes a portable iron church for the Diocese of Melbourne which contained nearly 700 seats, and the cost, including erection, was £1,000. The interior of a church for the Diocese of Melbourne is shown in a lithograph distributed by Hemming's firm (plate 161). The churches did not suit colonial conditions and only fragments remain.

St Stephen's Presbyterian Church, Sydney (plates 162–164) was imported from Glasgow and served four different purposes in its lifetime, church, lending branch of the Public Library, recreation hall and chapel of the Lidcombe State Hospital and Home, and finally scrap iron. The similarity to a church shown in *Illustrations of Iron Structures for Home and Abroad* manufactured by Charles D. Young and Co. (plate 163) suggests that this firm was the originator of St Stephen's Church[60].

Coppin's Royal Olympic Theatre in Melbourne, illustrated in *Victorian Heritage* (plate 48), was constructed by Messrs Bellhouse and Co. in England. Having a capacity of 1,150 persons, "this complete portable theatre of iron, with all fittings and appurtenances" cost a sum exceeding £4,000. The façade, faced with brick, was surprisingly graceful. However, the populace called the theatre the "Iron Pot," but whether derisively or affectionately is not known[61].

PLATE 159—St Luke's Church of England, Whitmore Square, Adelaide: the masonry front

PLATE 160—Looking towards the rear of St Luke's, showing the combination of stone, brick, and corrugated and sheet iron of which the body of the church was constructed. The front of the church may have been made of cast iron. The columns inside the church are wooden

PLATE 161—Lithograph showing the interior of a "Second Church erected for the Diocese of Melbourne at Hemming's Patent Improved Portable Buildings Manufactory" (by courtesy of the La Trobe collection, State Library of Victoria)

PLATE 162—A prefabricated iron building in Macquarie Street, Sydney, originally St Stephen's Presbyterian Church, and later the lending branch of the Public Library, photographed during demolition (by courtesy of the Mitchell Library, Sydney). The mode of construction of the roof and side wall of corrugated iron are well shown

PLATE 163—An illustration in a catalogue of Charles D. Young, of London, Liverpool, Glasgow, and Edinburgh. Similarities suggest that St Stephen's Church originated in the foundry of that firm. The Church had a pinnacle (or small tower) on each side of the façade, whereas that in the catalògue has a central one of similar general design. Other details differ, probably to give variety and an impression of uniqueness to other buildings sold. *(Reproduced by courtesy of the Royal Institute of British Architects, London)*

PLATE 164—Being an iron prefabricated building it was possible to put it together again in the grounds of the Lidcombe State Hospital. Drawing by A. Ashworth, 1956

PORTABLE DWELLINGS

A number of prefabricated dwellings may still be found in Melbourne. As might be expected, most are of the poorest type. A tiny house at present in Moor Street, Fitzroy, Victoria, was constructed by E. T. Bellhouse. Number 399 Coventry Street, South Melbourne, prefabricated by John Walker, is being restored by the National Trust of Australia (Victoria) at the suggestion of, and under the direction of, Mr Miles Lewis. A large and elegant house, Tintern, Toorak, Victoria, presents an iron façade visually indistinguishable from later additions of brick and stucco. The manufacturers were Messrs W. and P. McLellan of Glasgow. A number of firms obviously conducted a thriving trade in prefabricated buildings.

Recently the Baillieu Library, University of Melbourne, acquired a portfolio entitled *Hemming's Patent Improved Portable Houses*[62]. The houses were constructed of cast iron or timber framing, with walls and roof of galvanised corrugated iron, lined inside by half inch boarding covered with canvas, leaving a space of four and a half or three inches throughout the entire building between the iron and woodwork, to be later filled with any thermal non-conductor such as brickwork. Decorative castings were added. Included are a number of steel engravings and coloured lithographs, five of which depict buildings designed for Australian use (plates 165–169).

PLATE 165—Row of shops ordered for Melbourne; cost was 1,800 guineas

Plates 165–169 are copied from Hemming's Patent Improved Portable Houses[62], *Manufactory, Clift House, Bedminster, Bristol. By courtesy of the Baillieu Library, University of Melbourne*

PLATE 166—A house apparently bought by, or available from, the Melbourne and Colonial House Investment Company. Note the cast iron brackets. Mr Miles Lewis, architect, of Melbourne informs me that this house exists at Bowral, New South Wales

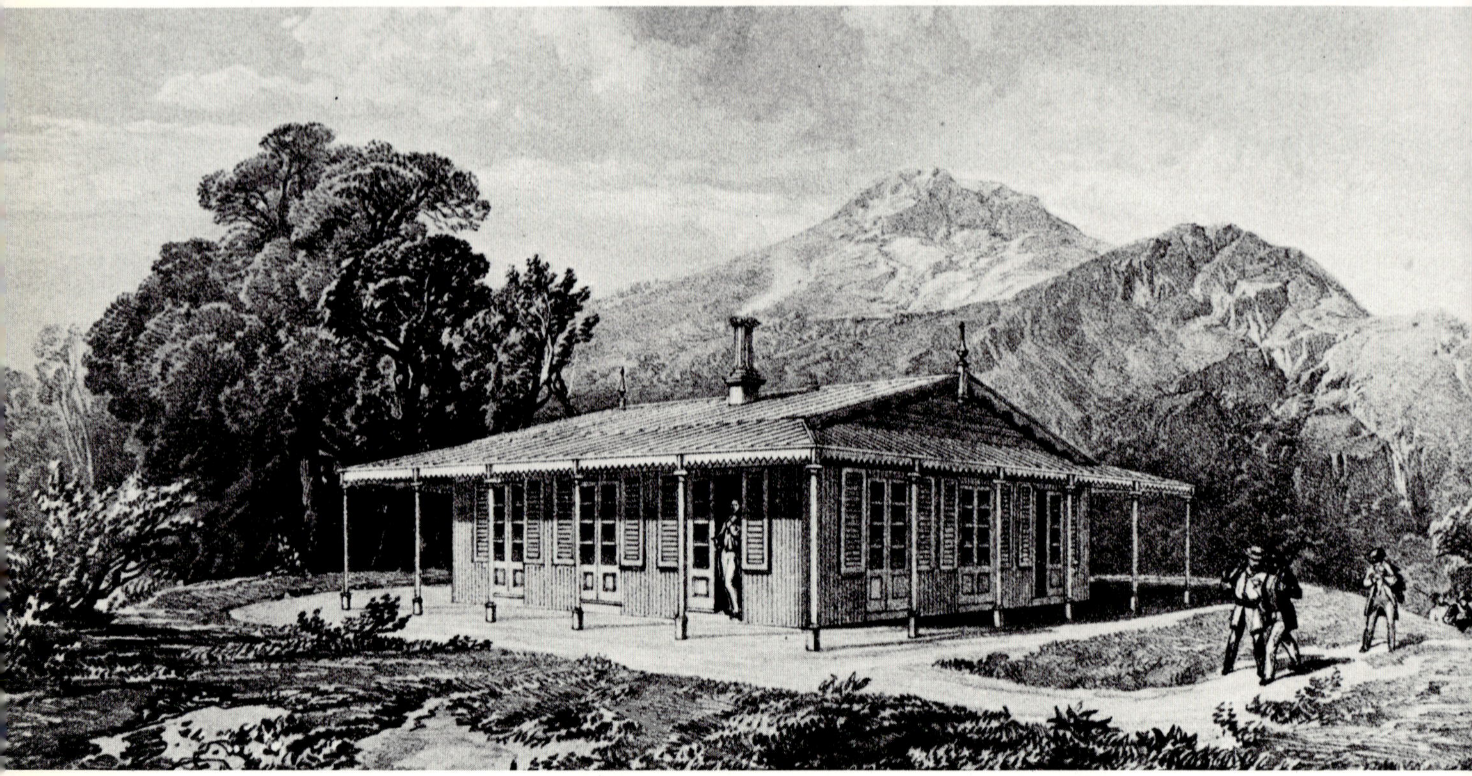

PLATE 167—Parsonage for the Diocese of Melbourne

PLATE 168—Proposed house for the Archbishop of Sydney

PLATE 169—Villa, to be erected at Sydney. Note the open-work columns and brackets

CORIO VILLA, GEELONG, VICTORIA

The most ornate of imported houses is Corio Villa, the exterior being made entirely of cast iron. In *Victorian Heritage* I quoted Mr G. E. Drinnan's architectural thesis dealing with the house and the only history known at that time—that is, packing cases containing the house were unloaded at a wharf in Geelong in 1855, where it remained unclaimed. Attempts to find the owner failed: enquiries directed to the consignor elicited the information that the foundry had been burnt down and all its records destroyed. The shipment was eventually sold to Mr Alfred Douglass, who succeeded in having it assembled without plans (plates 170–173).

While in Edinburgh, on 29 September 1969, I met Mr Malcolm Higgs of the Department of Architecture, University of Edinburgh, who was writing a thesis on prefabrication. He brought along a copy of *Illustrations of Iron Structures for Home and Abroad* manufactured by Charles D. Young and Company of Edinburgh[63] to enquire whether figure 14 (plate 174 in this book) corresponded with the Corio Villa which I had illustrated in *Ornamental Cast Iron in Melbourne*[64].

The legend was as follows:

> Figure 14 is referred to as an excellent example of a Country Villa, in a neat Style of Architecture, with Verandas to correspond. The Villa, from which this is a drawing as actually erected, was made for the late Mr Gray, Colonial Land Commissioner at Geelong.

Knowing Corio Villa well I was able to confirm Mr Higgs's suspicion.

It will be noted that there are two major differences—the placement of one of the elaborately decorative bowed verandahs and the design of the straight frontal colonnade. Such variations would be easily effected by the method of construction and ordering. Mr M. McAllister, the present owner of Corio Villa, is painstakingly and carefully restoring the villa to its original state, and has in mind the removal of the false wall between the columns which will reveal the verandah and its iron fenestrated wall shown in figure 174.

The cast iron vases in the garden are identical with those shown in a catalogue of the same firm entitled *Illustrated and Descriptive Catalogue of Ornamental Cast and Wrought Iron and Wire Work* dated 1850, which I have had for some years. The vases on the bowed verandahs are of the same form as the Medici vases shown on page 48 of the catalogue[65]. The figures, however, are different (plate 175). Urns in the garden are exactly the same as one shown in the catalogue (plate 176)[66].

The prices quoted in the catalogue were:

30 inches high,	with	figures, exactly	as shown above	£6 6s.	each
19	,,	,, do.	do.	£3 3s.	,,
30	,,	,, without figures	do.	£5 5s.	,,
19	,,	,, do.	do.	£2 10s.	,,

It seemed highly probable that Mr Gray, Colonial Land Commissioner at

Geelong, was William Nairn Gray, for *Blue Books* (which list civil servants) held in the Archives Division of the La Trobe Library, Melbourne, show only one Commissioner of Crown Lands named Gray[67]. He was appointed Commissioner of Portland Bay District on 1 July 1851. In 1853 he was, in addition, Police Magistrate. He was not mentioned in the 1854 *Blue Book,* but his death notice appeared in *The Argus* of 17 June 1854, and this revealed that he died at Hamilton on 11 June 1854 (his second name was spelt Nairne). It is little wonder that he failed to claim his villa, which arrived in 1855.

Mr P. L. Brown, immediate past president of the Geelong Historical Society kindly added the following information of William Nairn Gray.

Historical Records of Australia (Vol. 17, page 409) reported the appointment of W. N. Gray, Esqr., as Police Magistrate at Illawarra by Governor Bourke on 15 April, 1834. Further reports show that he served as magistrate at Port Macquarie, until he was displaced by reduction of estimates in 1847 *(Historical Records of Australia,* Vol. 26, page 167). *The Port Phillip Gazette* of 9 May 1849 (page 302) states that on 25 April 1849 'His Excellency the Governor has been pleased to appoint William Nairn Gray, Esquire, to be Police Magistrate for the Town of Geelong.'

Mrs W. G. Jones (née Claire Douglass) of Ocean Grove, is the granddaughter of Alfred Douglass and lived at Corio Villa as one of H. P. Douglass's children.

Figure 15 from Young's *Illustration of Iron Structures* (shown here as plate 177) carries the following description:

> Figure 15 was constructed for Messrs. Miller and Dismor, of Melbourne, and erected in Collins Street there. The greater portion is of cast iron, and it was the first building of its kind sent out to Australia. It was made so far back as 1851, and shows to what perfection the system had even then attained[68].

The title page of Young's catalogue is reproduced in plate 178.

PLATE 170—Corio Villa, Geelong

PLATE 171—The side verandah of Corio Villa

Right

PLATE 172—Detail of the columns, hollow and square in section, Corio Villa

Below

PLATE 173—The front of Corio Villa on the right-hand side of the entrance door

PLATE 174—Charles D. Young's country villa.[64]. Some changes were made in the actual building from that shown in the catalogue. One of the bowed verandahs has been changed in position and square columns have been substituted for round columns along the front of the house. A wooden portico between the two bows was added locally, and at some stage the arcades along the front of the house have been filled in. There can be no doubt that Corio Villa was cast by Charles D. Young. *(Reproduced by courtesy of the Royal Institute of British Architects)*

PLATE 175—A study of the urns removed any doubt of the origin of Corio Villa. The Medici urns on the bowed verandahs have exactly the same form as those in Young's *Illustrated and Descriptive Catalogue of Cast and Wrought Iron and Wire Work*[65] , dated 1850

THE MEDICI VASES.

'othing can exceed the beauty of these examples: they are quite equal ıe great originals, in proportion, execution, sharpness, and beauty of ne. The figures stand in bold relief, and as specimens of Iron Casting are unequalled in the world.

PLATE 176—A garden urn, Corio Villa, showing its similarity to an urn in Young's catalogue[66]

No. 1.

PLATE 177—A shop shown in Young's *Illustrations of Iron Structures for Home and Abroad,* erected in Collins Street for Messrs Miller and Dismor[68]. "It is stated to be one of the handsomest shops in Melbourne." *(Reproduced by permission of the Royal Institute of British Architects, London)*

ILLUSTRATIONS

OF

IRON STRUCTURES,

FOR

Home and Abroad,

CONSISTING OF

STORES, DWELLING-HOUSES, MARKETS,

ARCADES, RAILWAY STATIONS, AND ROOFINGS, &c. &c.,

CONSTRUCTED OF WROUGHT AND CAST IRON AND

CORRUGATED SHEETS,

MANUFACTURED BY

Charles D. Young & Company,

19 GREAT GEORGE STREET, WESTMINSTER..........LONDON.
1 CASTLE BUILDINGS, DERBY SQUARE..................LIVERPOOL.
32 ST ENOCH SQUARE..GLASGOW.
48 NEW BUILDINGS, NORTH BRIDGE........................EDINBURGH.

EDINBURGH.

PLATE 178—Title page of Young's catalogue *(Reproduced by permission of the Royal Institute of British Architects, London)*

4

ENGLISH REGISTRATION MARKS

IRON FABRICATIONS were registered in England in the same manner as ceramics, wood, and glass objects. The registration mark of a hall stand of Gothic design (plate 179) is shown in plate 180.

The registration mark was used between 1842 and 1883. The material was indicated (metal, wood, glass, and ceramics) by the Roman numerals I to IV in the circle at the top of the mark. Up to 1868 the letter below the class number gave the year of manufacture. From 1868 to 1883 the figure below the class number is the day of the month of manufacture, while the month is indicated by the letter at the bottom. The letter on the right corner of the diamond gives the year of manufacture. The figure opposite it is the manufacturer's identification. Unfortunately enquiry to the British Patents Office failed to produce information of the manufacturers' designations.

Before the year 1868 the day of the month is recorded in the right hand corner, while the month is indicated in the left hand corner. The figure at the bottom indicates the manufacturer, who may be recorded separately as in figure 181.

The Antique Dealer and Collectors' Guide provides the following information:

YEAR LETTERS

1842	X	1856	L	1870	C
43	H	57	K	71	A
44	C	58	B	72	I
45	A	59	M	73	F
46	I	60	Z	74	U
47	F	61	R	75	S
48	U	62	O	76	V
49	S	63	G	77	P
50	V	64	N	78	D
51	P	65	W	79	Y
52	D	66	Q	80	J
53	Y	67	T	81	E
54	J	68	X	82	L
55	E	69	H	83	K

MONTH LETTERS

January	C	May	E	September	D
February	G	June	M	October	B
March	W	July	I	November	K
April	H	August	R	December	A

PLATE 179—A hallstand of Gothic design

PLATE 180—The registration mark on the back of the hallstand, indicating 31 March 1873 as the date of manufacture in England

PLATE 181—Almost ideal identification: a door-knocker manufactured by A. Kenrick and Sons. However, details of the date are difficult to decipher (probably 12 January 1854)

5

MERCHANDISING

FOUNDRIES MAY HAVE SOLD their products, and perhaps those of other foundries, direct to builders, but some, at least, were sold by other merchants. Mr L. A. Daniel of Fulham, South Australia, was kind enough to present me with a catalogue of Metters Limited[69]. Mr Daniel had retired after being with the firm for fifty-two years. The catalogue was not dated, but the typography suggests that it was produced early in the present century (plate 182).

The catalogue of forty-eight pages was concerned chiefly with ornamental castings. It is of interest that while only three pages showed balusters of twenty patterns, forty-six designs of matching brackets and friezes occupied twenty-three pages (plates 183–186). This probably reflected a trend towards single storey houses at that time. A number of patterns illustrated appeared in the catalogues of Adelaide and Melbourne foundries, and some were registered in other States. Two pages were devoted to cast iron Garden Gates and Railings (plate 187).

Mr Daniel, who began working with Metters in 1918, recalls the early history of the business. "Mr Fred Metters, son of an English migrant, established a foundry in Adelaide in 1891 for the manufacture of fuel stoves. He toured country shows in South Australia demonstrating the stoves, and as the business prospered he opened an office and showrooms in Rundle Street, Adelaide, and later branches were opened in Perth, Sydney, and Melbourne. In addition to other items, he manufactured those shown in the catalogue—cast iron balusters, brackets, friezes, fringes, columns, grave tablets, gates, and fences. In 1907 he sold his interests to the company which was formed with a nominal capital of £200,000 in Adelaide. The ornamental castings were discontinued from manufacture about 1920, and thereafter the firm produced baths, basins, sinks, fuel, gas, and later electric stoves, heating stoves, grates, agricultural equipment, windmills, tanks, and kitchenware.

"In 1927 Metters amalgamated with K.F.B. Foundry Ltd, and a holding company, General Industries, was formed. Metters were iron-founders, wholesalers, and retailers, although the latter was only a small fraction of the firm's business. Most of the business was obtained from hardware merchants."

PLATE 182—The cover of a *Metters Limited* catalogue

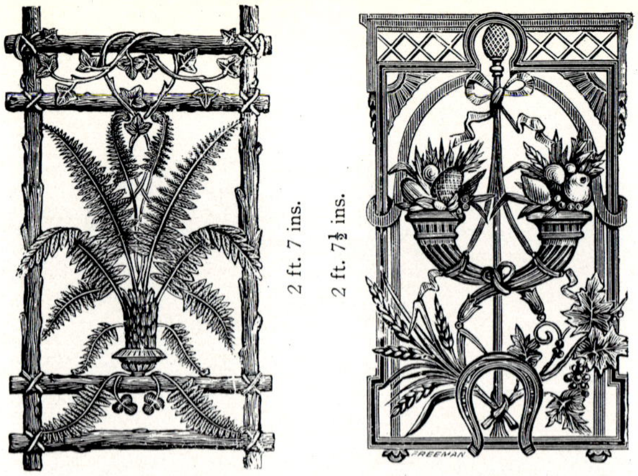

No. 142. No. 144.

Plate 183—Two balusters from Metters catalogue. (No. 142 is also to be found in H.C. as No. 580, plate 9 in this book. It was also registered in Victoria by Cochrane and Scott, No. 335, 5.8.1884. No. 144 was registered in Victoria by the same firm—Vic. 635, 7.4.1892. It also appears in H.C., No. 580)

Plate 184—Brackets, etc., Metters catalogue. (No. 1, p. 2. The pattern is also in H.C., No. 940A, plate 13 in this book. The bracket and frieze are shown with a slight modification as No. 33 in the catalogue of William Stephens's Excelsior Foundry, South Melbourne[10])

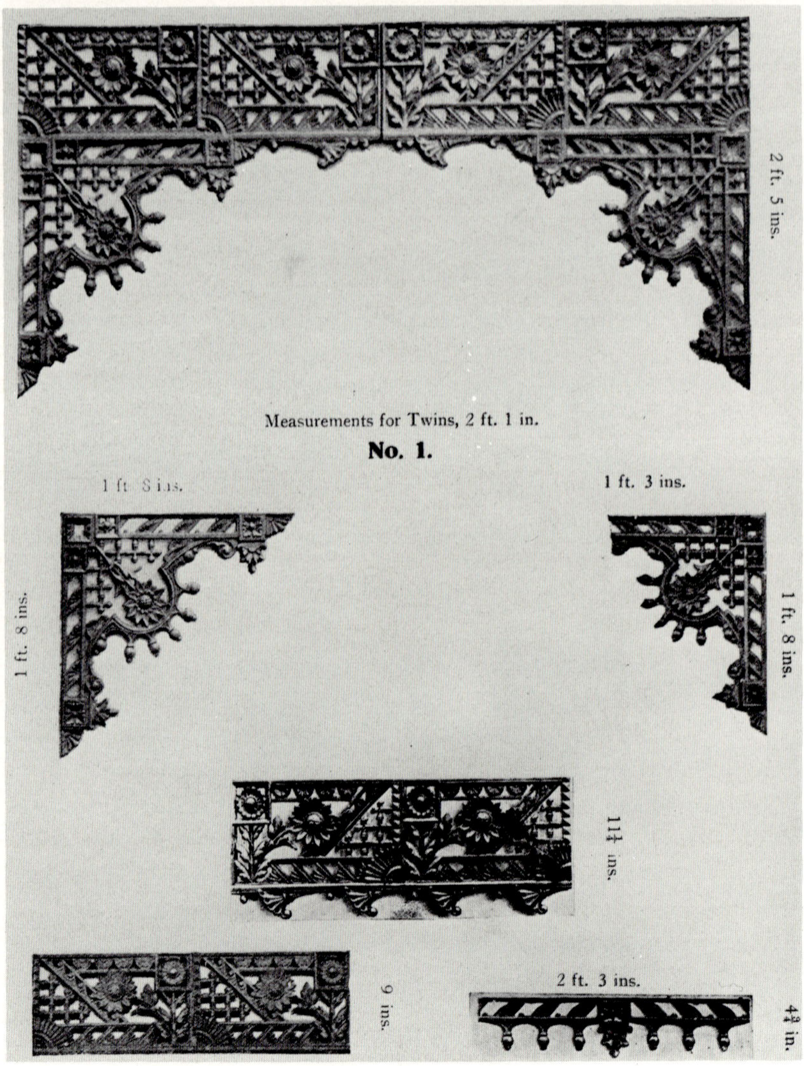

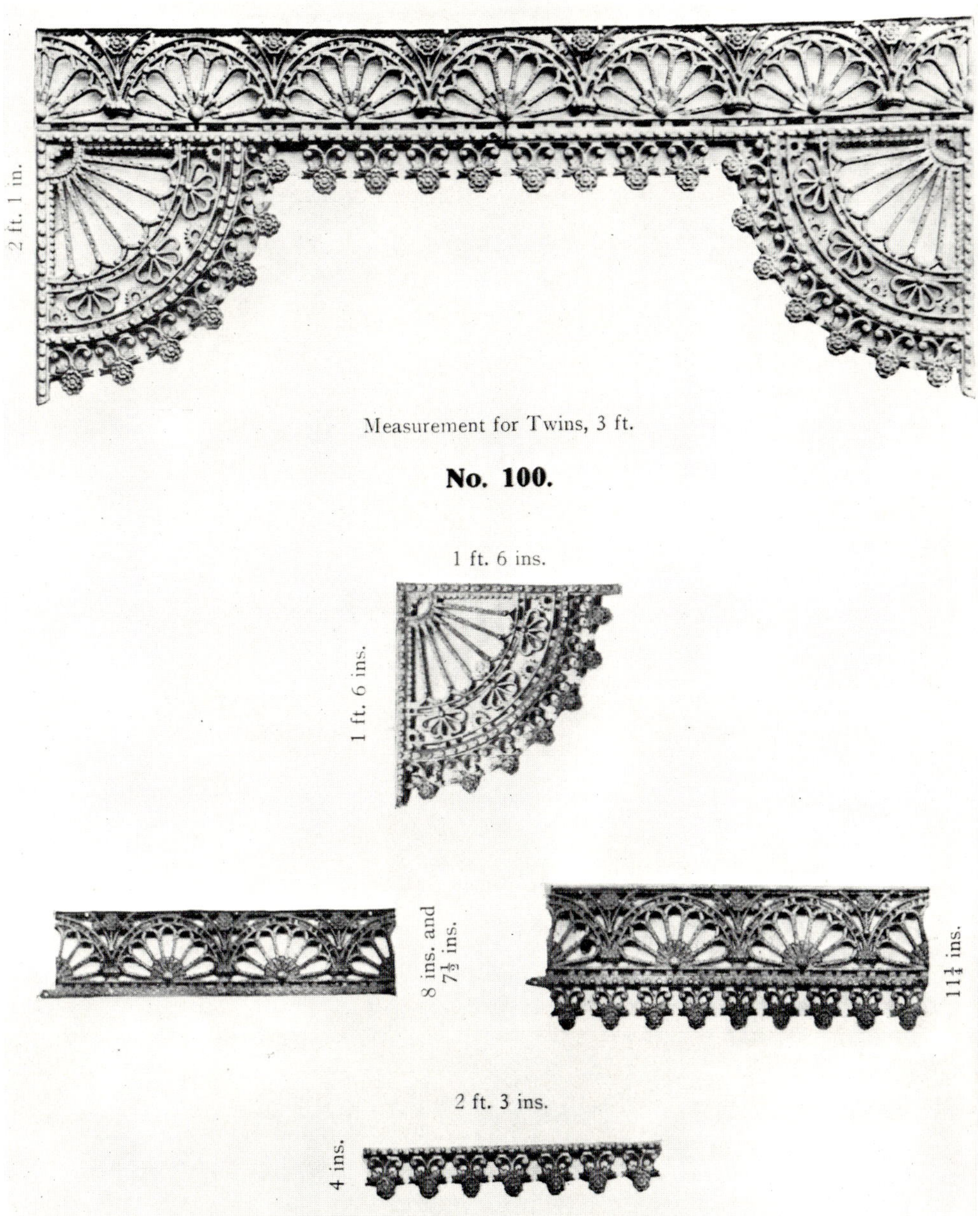

PLATE 185—Brackets, etc., Metters catalogue. (No. 100, p. 16, also H.C., No. 418A, plate 12 in this book and the Stephens's catalogue, No. 29A)

PLATE 186—Brackets, etc., Metters catalogue. (No. 158, p. 19; also shown in the catalogue of the Excelsior Foundry of William Stephens, South Melbourne, No. 34, p. 17)

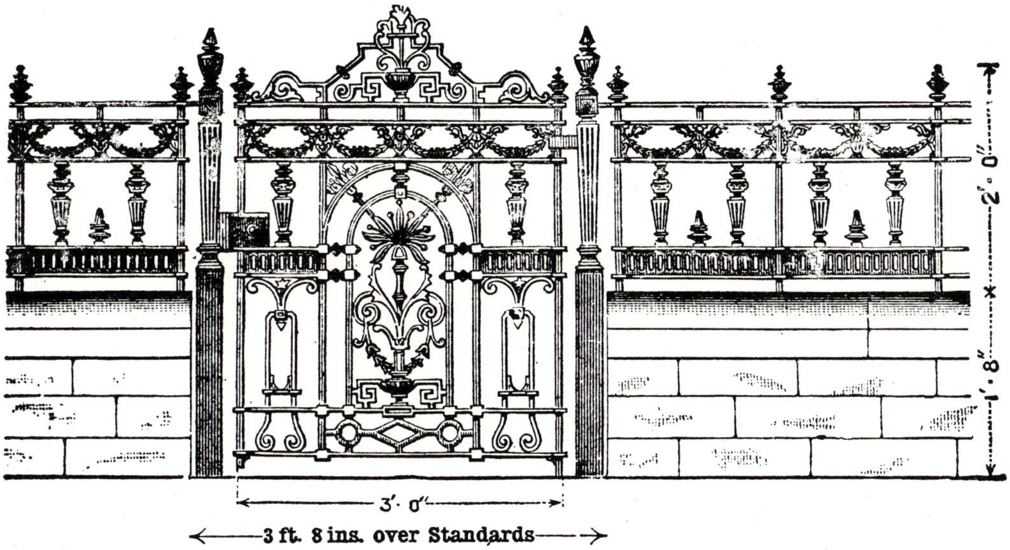

No. 91.

No. 80

Plates 187a *(above)* and 187b *(overleaf)*—Cast iron garden gates and railings, Metters catalogue (pp. 34, 35). Two pages of the catalogue deal with this feature of cast iron ornamentation, so characteristic of Adelaide

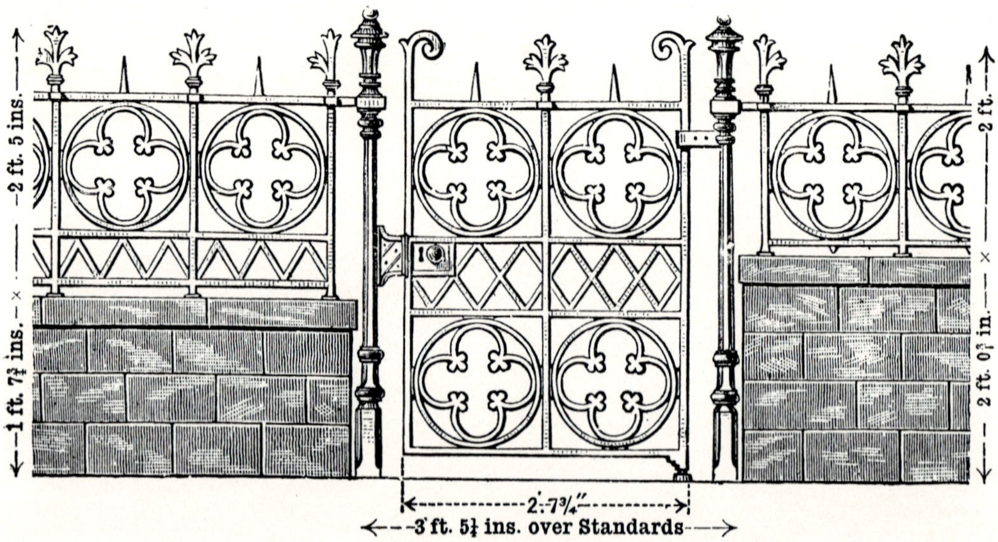

No. 20 Railing. No. 21 Gate. No. 22 Railing.

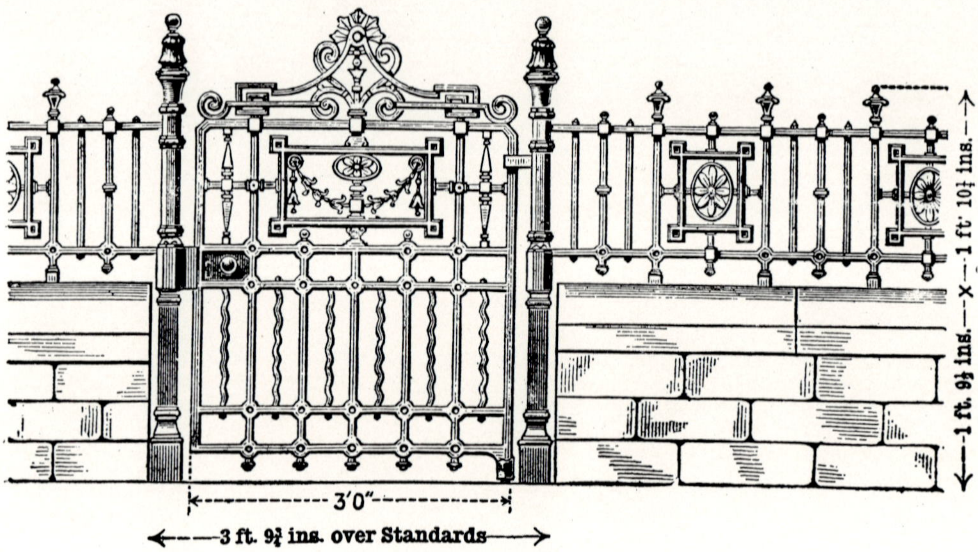

No. 89.

PLATE 187B

EPILOGUE

As a final exercise in nostalgia, plate 188 records the prices of decorative architectural ironwork obtaining at the turn of the century[70].

FRIEZES AND BRACKETS.

Plate No.	No.		FRIEZE Per Foot (lineal).	BRACKETS. Each
1	1	Crown	1/6	3/3
"	2	Fern...	1/3	2/9
"	3	New Scroll	11d.	2/3
"	4	Trumpet	1/2	2/9
"	5	Shield	8d.	2/9
2	6	Daisy	1/4	2/6
"	7	Fruit...	11d.	2/3
"	8	New Ring	11d.	2/3
"	9	Star ...	8d.	2/6
"	10	Shield	1/6 14"	2/9
"	—	Shield	1/3 12"	—
3	11	Bird ...	1/3	3/-
"	12	Dolphin	1/-	2/3
"	13	Acorn	1/-	2/3
"	14	Scroll	1/-	2/3
4	15	Centennial	1/3 12"	2/3
"	—	Centennial	1/- 10"	2/3
"	16	Trumpet	10d.	2/9
"	17	Scroll	11d.	2/3
"	—	Scroll	9d.	2/3
"	18	Centennial	9d.	2/3
"	19	Scroll	8d.	2/3
5	20	Sunflower	1/3 12"	2/9
"	—	Sunflower	1/6 14"	3/9
"	21	Heart	9d.	2/-
"	22	Athol	11d.	2/-
"	23	Gothic	1/5	2/6
6	24	Heart	1/2	2/3
"	25	Star ...	6d.	2/6
"	26	Crown	9d.	2/3
7	27	Dolphin	1/6	2/6
"	28	New Ring	1/5	2/9
15	29	Fan ...	1/2	3/-
"	30	Daisy	1/-	2/9
"	—	Daisy	1/6	3/9
"	31	Athol	10d.	2/3
16	32	Star Scroll	10d.	2/3
"	33	Æsthetic	1/6	2/9
"	—	Æsthetic	2/-	3/9
17	34	Holly	1/3	2/9
"	35	Leaf...	1/-	2/9
"	36	Gothic	1/1	2/3
"	37	Leaf...	1/3	2/9
"	38	Gothic	11d.	2/-
21	39	Sunbeam	1/3	2/6
"	40	Daisy Ring	1/2	2/6
"	41	Star Scroll	11d.	2/6
"	42	Diamond	1/2	2/9
"	43	Palm...	1/3	2/6
"	44	New Scroll	1/3 12"	2/3
"	—	New Scroll	1/6 14"	2/6
"	45	Star Scroll	1/3	2/6
"	46	Oriental	1/3	2/6

CRESTING.

Plate No.	No.	
6	29	per foot 1/9, 3d. per foot extra, with Clips.
"	30	" " 2/-, " " " " " "
"	31	" " 1/6, " " " " " "
24	32	" 14" 1/-, " " " " " "
"	—	" 21" 1/6, " " " " " "

BALCONY PANELS.

Plate No.	No.	Per Foot (lineal).	Per Sheet.
7	1	3/-	—
"	2	2/10	—
8	3	3/-	—
"	4	— 20" wide, 4/6	
"	5	3/-	—
"	6	—	5/-
"	7	3/-	—
"	8	3/-	—
"	1	Sunflower, Standard, 1/3 each.	
"	2	Daisy " 1/3 each.	
9	9	3/-	—
"	10	3/-	—
"	11	3/-	—
"	12	3/-	—
"	13	3/-	—
"	14	3/3	—
15	15	—	5/6
"	16	3/-	—
16	17	2/9	—
"	18	3/-	—
17	19	3/-	—
"	20	—	5/6
"	21	3/-	—
"	22	—	4/3
23	23	3/-	—
"	24	3/-	—
"	25	—	4/9
"	26	4/-	—
"	27	—	3/6
"	28	—	3/6
"	29	3/-	—
"	30	3/-	—

SPANDRILS.

Plate No.	No. of Article.	Each.	Plate No.	No. of Article.	Each.
11	1	7/–	11	6	7/–
"	2	7/–	"	7	5/–
"	3	6/–	"	8	5/–
"	4	6/–	"	9	5/–
"	5	5/–	"	10	4/–

FRIEZE PANELS.

Plate No.	No.	Per Lineal Foot.	Plate No.	No.	Per Lineal Foot.
10	1	1/–	18	24	10d.
"	2	1/3	"	25	10d.
"	3	1/–	"	26	10d.
"	4	11d.	"	27	1/–
"	5	11d.	"	28	11d.
"	6	1/1	"	29	1/–
"	7	1/–	"	30	11d.
"	8	11d.	22	31 Melb. S. Corp.	1/–
"	9	11d.	"	32	1/3
"	10	11d.	"	33	1/3
"	11	11d.	"	34	1/–
"	12	1/1	"	35	11d.
18	13	1/3	"	36	10d.
"	14	11d.	"	37	11d.
"	15	9d.	"	38	1/4
"	16	1/2	"	39	9d.
"	17	11d.	"	40	9d.
"	18	10d.	"	41 Prahran Corp.	1/–
"	19	10d.	"	42	1/3
"	20	6d.	"	43	1/–
"	20	10d.	"	44	1/–
"	20	11d.	"	45	10d.
"	21	6d.	"	46	10d.
"	22	1/–	"	47	10d.
"	23	10d.			
"	23	1/–			

FRINGES.

Plate No.	No.	Per Lineal Foot.			Price.
12	1	8d.	Brackets to match		2/9
"	2	8d.	"	"	2/9
"	3	6d.	"	"	2/6
"	4	6d.	"	"	2/6
"	5	5d.	"	"	2/6
"	6	6d.	"	"	2/9
"	7	7d.	"	"	4/–
"	8	6d.	"	"	2/9
"	9	4d.	"	"	2/6
"	10	4d.	"	"	2/6
"	11	5d.	"	"	2/9
24	11A	8d.	"	"	2/9
"	12	9d.	"	"	2/9
"	13	7d.	"	"	2/9
"	14	8d.	"	"	2/9
"	15	6d.	"	"	2/9
"	16	6d.	"	"	2/9
"	17	8d.	"	"	—
"	18	7d.	"	"	—

Nos. 5, 6, 8, 11A, 13, 14 and 17 make good Crestings.

PLATE 188—Price list from the catalogue of William Stephen's Excelsior Foundry, South Melbourne, showing the prices obtaining at the turn of the century

BIBLIOGRAPHY

The numbers before each entry are those used as references in the text.

1. Maass, John. *The Gingerbread Age*. New York: Rinehart & Co., 1957.
2. Robertson, E. G. *Victorian Heritage*. Melbourne: Georgian House, 1960.
3. ———. *Sydney Lace*. Melbourne: Georgian House, 1962.
4. ———. *Ornamental Cast Iron in Melbourne*. Melbourne: Georgian House, 1967.
5. Feininger, Andreas. *Man and Stone*. New York: Crown, 1961.
6. Mortensen, William. *The Command to Look*. San Francisco: Camera Craft, 1937.
7. Brandt, Bill. Quoted by John Szarkowski, as an introduction to an exhibition of Brandt's photography at the Museum of Modern Art, New York.
8. Jose, A. W., and Carter, H. J., eds. *The Australian Encyclopaedia*. 3rd ed., rev. Sydney: Angus and Robertson, 1927. p. 72.
9. Robertson, E. G. and Craig, E. N. *Early Houses of Northern Tasmania*. Melbourne: Georgian House, 1964.
10. Excelsior Foundry. *Illustrated Catalogue*. South Melbourne: William Stephens, n.d.
11. Robertson, E. G. *Ornamental Cast Iron in Melbourne*, pp. 44, 48–63.
12. Stewart & Harley. *"Sun" Foundry, Illustrated Catalogue*. Adelaide: Crescent Printery, 1897.
13. Harley, A. C. & Co. *"Sun" Foundry, Illustrated Catalogue*. 2nd ed. Adelaide: Vardon & Sons, 1914.
14. Fulton, G. E., & Co. *Illustrated Catalogue of Fulton's Castings*. 2nd ed. Adelaide: J. H. Sherring & Co., 1887.
15. Robertson, E. G. *Ornamental Cast Iron in Melbourne*, pp. 19–25.
16. Robertson, E. G. *Sydney Lace*, pp. 19–25.
17. Morgan, E. J. R. and Gilbert, S. H. *Early Adelaide Architecture*. Melbourne: Oxford University Press, 1969. p. 23.
18. Henn, L., & Co. *A Series of Views of South Australia*. Adelaide: n.d. (c. 1879).
19. Morgan, E. J. R. and Gilbert, S. H. *Early Adelaide Architecture*, p. 27.
20. Ibid., p. 8.
21. Gill, S. T., Nixon, F. R., Calvert, S., and Korn, O. *Fourteen Views of Old Adelaide*. Adelaide: E. S. Wigg & Son, n.d. (c. 1889).
22. Harcus, William. *South Australia: its History, Resources and Productions*. London: Sampson Low, Marston, Searle and Rivington, 1876. p. 38.
23. Morgan, E. J. R. and Gilbert, S. H. *Early Adelaide Architecture*, p. 38.
24. Robertson, E. G. *Sydney Lace*, p. 20.
25. Morgan, E. J. R. and Gilbert, S. H. *Early Adelaide Architecture*, p. 102.
26. Robertson, E. G. *Ornamental Cast Iron in Melbourne*, p. 41.
27. Morgan, E. J. R. and Gilbert, S. H. *Early Adelaide Architecture*, p. 32.
28. Macfarlane, Walter & Co. *Illustrated Catalogue of Macfarlane's Castings, Saracen Foundry*. 6th ed. Glasgow: n.d. p. 339.
29. Robertson, E. G. *Sydney Lace*, plate 208, p. 183.
30. *The Australian Town and Country Journal*. Volume II, No. 27 (9 July 1870).

31. MACFARLANE. *Illustrated Catalogue.* 6th ed.

32. SMITH, GEO. & CO. *Illustrated Catalogue of the Sun Foundry.* Glasgow: n.d.

33. McDOWALL, STEVEN & CO. *Catalogue.* 21st ed. Section III. Specimen Plate, Balusters, No. 507.

34. MACFARLANE. *Illustrated Catalogue.* 6th ed. Preface.

35. *Encyclopaedia Britannica.* 13th ed. London: The Encyclopaedia Britannica Co. Ltd., 1926. Volume 27, p. 128.

36. *The Australian Ironmonger.* Volume II, No. 4 (1 April 1887). pp. 84, 85.

37. COTTINGHAM, L. N. *The Ornamental Metal Worker's Director.* 1823. Preface.

38. ROBERTSON, E. G. and CRAIG, E. N. *Early Houses of Northern Tasmania,* Volume 1, p. 144.

39. COTTINGHAM, *The Smith and Founder's Director,* 1824, plate I, No. 1.

40. Ibid, plate II, No. 8.

41. ROBERTSON, E. G. *Sydney Lace,* plate 1.

42. LOYAU, G. E. *The Gawler Handbook.* Adelaide: Goodfellow & Hele, 1880. p. 52.

43. Ibid., p. 53.

44. Ibid., p. 14.

45. COOMBE, E. H. *History of Gawler, 1837–1903.* Adelaide: Vardon & Sons, 1910. p. 351.

46. Ibid., p. 354.

47. Mittagong Iron Week Celebrations Committee. Souvenir.

48. National Trust of Australia (Victoria). *Trust Newsletter.* September 1967. p. 1. (A definitive study of Lal-Lal is being made by Mr Graeme Cope, Monash Teachers' College, North Clayton, Victoria.)

49. FOWLES, JOSEPH. *Sydney in 1848.* Sydney: J. Fowles, p. 55.

50. University of Sydney. *The Gazette.* Volume 2, No. 17 (April 1970). p. 298.

51. FOWLES. *Sydney in 1848,* facing p. 38.

52. Ibid., p. 39.

53. ROBERTSON, E. G. *Sydney Lace,* pp. 41, 69, 84, 89, 100, 104, 187.

54. BAKER, R. T. *The Australian Flora in Applied Art.* Sydney: 1915.

55. SMITH, J. E. *A Specimen of the Botany of New Holland.* London: J. Sowerby, 1793. p. 19.

56. WHITE, JOHN. *Journal of a Voyage to New South Wales.* London: J. Debrett, 1879.

57. ROBERTSON, E. G. *Ornamental Cast Iron in Melbourne,* p. 58.

58. MACFARLANE. *Illustrated Catalogue,* Volume 2, 6th ed. Glasgow: 1882. p. 674.

59. *St Luke's Church, 1855–1955.* Centenary Booklet. Adelaide: Hunkin, Ellis & King Ltd., 1955. p. 1.

60. YOUNG, CHARLES D., & CO. *Illustrations of Iron Structures for Home and Abroad.* Edinburgh: n.d.

61. ROBERTSON, E. G. *Victorian Heritage,* pp. 44–45.

62. *Hemming's Patent Improved Portable Houses.* Sole Manufactory, Clift House, Bedminster, Bristol. n.d.

63. YOUNG, CHARLES D. *Illustrations of Iron Structures for Home and Abroad,* figure 14.

64. ROBERTSON, E. G. *Ornamental Cast Iron in Melbourne,* p. 179.

65. YOUNG, CHARLES D., & CO. *Illustrated and Descriptive Catalogue of Ornamental Cast and Wrought Iron and Wire Work.* Edinburgh: 1850. p. 48.

66. Ibid., p. 49.

67. Information supplied by Miss Pat Reynolds, Archivist, La Trobe Library, Melbourne.

68. YOUNG, CHARLES D. *Illustrations of Iron Structures for Home and Abroad.*

69. METTERS LIMITED. *Metters Limited.* North Adelaide: Donald Taylor Collotype Co., n.d.

70. EXCELSIOR FOUNDRY. *Illustrated Catalogue,* Price Lists.

INDEX

Plate numbers are shown in bold type